PROJECT APOLLO
The Way to the Moon

PROJECT APOLLO
The Way to the Moon

by

P. J. BOOKER
G. C. FREWER
G. K. C. PARDOE

American Elsevier Publishing Company, Inc.
New York 1969

American Edition Published by
American Elsevier Publishing Co., Inc.
52 Vanderbilt Avenue, New York, NY10017

Standard Book Number 444-19705-2

Library of Congress Catalog Card Number 72-101222

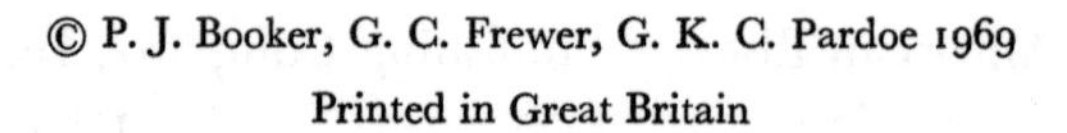

Printed in Great Britain

CONTENTS

ILLUSTRATIONS

The illustrations appear between pages 72 & 73

FOREWORD

'It is difficult to say what is impossible, for the dream of yesterday is the hope of today and the reality of tomorrow.'

Dr. Robert Goddard, the Father of American Rocketry.

PROJECT APOLLO will, in due course, affect every living person on the Earth, for through it some of our kind have, for the first time in man's long history, travelled out into the universe and set foot on another heavenly body. The fact that this body is but the Moon, a natural satellite of the Earth familiar to us all and only on the inner fringe of outer space, does not detract from the significance of the event, truly a landmark in the story of human progress.

People of different countries and cultures, with varying outlooks and needs, will view this excursion to the Moon in different ways–but whatever their viewpoints, none can deny that Project Apollo represents a stupendous technical achievement, and this is what this book is about.

It is a straightforward account of how the engineers and technologists tackled the problem of how to put men onto the Moon, and how they carried the operation through successfully. The viewpoint is that of three engineers–two British and one British-born American–each with a different type of involvement and interest in the project and with a different kind of contribution. Our viewpoint is, therefore, Anglo-American–while the honours must go to America, who has planned and financed the project and driven it through successfully, nevertheless this has been an achievement of all mankind, the culmination of thought and action stretching over many generations in Britain, France, Germany, Russia and many other countries as well as America.

Others have written about the astronauts and their courage, about the political and management situations inseparable from such a vast operation, and about the scientific aims of the project. Here we seek to present the achievement mainly in technological terms, but without jargon or mathematics. The reader seeking just an overall picture can gloss over many of the facts and figures provided without losing the thread, while the more professional reader will, it is hoped, find sufficient information to appreciate some of the finer points of the engineering tasks.

No attempt is made to justify going to the Moon. In the end one either accepts that man's destiny lies this way or one does not. But the fact remains that a relatively small band of men with vision and dedication have pushed the world into a new age, making the inevitable happen now rather than in the next generation. Project Apollo is a mighty achievement, and we should be proud that it has happened in our time.

Peter Booker,
Gerald Frewer,
Geoffrey Pardoe

August 1969

ACKNOWLEDGEMENTS

The authors are pleased to acknowledge the very generous assistance provided by the National Aeronautics and Space Administration, directly through the Public Relations Departments at both Washington and the Kennedy Space Center, and indirectly through the United States Information Service in London; also that provided through the Public Relations Offices of the major contractors, The Boeing Company, North American Rockwell Corporation, Grumman Aircraft Engineering Corporation, McDonnell Douglas Company and the Chrysler Corporation; as well as the help given by engineers of the Boeing Atlantic Test Center and others at the Kennedy Space Center complex.

Although the majority of the line illustrations have been drawn specially for this book, they are based on original material provided by one or more of the above agencies and companies.

The photographs for plates 6A and 6B, and Figs. 20 and 21 were provided by the North American Rockwell Corporation Space Division; Fig. 14 was provided by the Grumman Aircraft Engineering Corporation. All other photographs are from the archives of the National Aeronautics and Space Administration.

1
First Steps in Manned Space Flight

IN May 1961, before a Joint Session of the House and Senate of the United States of America, President John F. Kennedy told the world that the USA had set itself the goal of landing a man on the Moon and returning him safely to Earth in the following decade. 'No single space project in this period,' he said, 'will be more exciting, or more impressive to mankind, or more important for the long range exploration of space; and none will be so difficult or expensive to accomplish.' He had no illusions over the magnitude of the task and he continued, 'But in a very real sense, it will not be one man going to the Moon–we make this judgment affirmatively–it will be an entire nation. For all of us must work to put him there . . . money alone will not do the job. This decision commands a major national commitment of scientific and technical manpower, material and facilities. . . . It means a degree of dedication, organisation and discipline, which have not always characterised our research and development efforts . . . that this Nation will move forward, with the full speed of freedom, in the exciting adventure of space.'

It is important to recognise that the American aim was not and is not just that of getting a man on the Moon. The objective was the creation of a viable space capability, to be attained by building up the technology, know-how, hardware, administration and operational experience upon which future space initiatives would be based. Landing a man on the Moon was a project which would create this capability, and one which was understandable and challenging enough to command the support that would be needed over the following years of endeavour.

The project was given the name *Apollo*, appropriately enough after the Greek god of light and the twin brother of

Artemis, the goddess of the Moon, and it received the 'go ahead' in 1961, though this is not really the beginning of the story. A momentous decision like this could not be taken unless there was sufficient evidence available to show that such a course was feasible and had a high chance of success. This decision was, in fact, the culmination of a succession of events spread over many years of history.

The Moon, being our nearest neighbour–a mere quarter of a million miles away–has tantalised men of vision for centuries. Indeed, without this target on our doorstep, as it were, it is difficult to conceive man seriously starting on the road to space exploration. Jules Verne and H. G. Wells wrote famous fictional books about trips to the Moon, though the methods they used to get their passengers space-borne were hardly practical, relying either on giant guns or an 'anti-gravity' material. Although it took time for his ideas to spread, it was the Russian mathematician Konstantin Tsiolkovsky who showed that the only successful way for man to explore the universe was by way of rockets, missiles which could attain very high speeds in one direction by ejecting part of their own mass–usually in the form of high-speed gases resulting from burning fuel–in the opposite direction.

The possibility of space travel within the solar system led to the formation of Interplanetary Societies in many countries, and much useful theoretical work was carried out by members of these bodies. There was, however, little real progress until it was suggested that as part of the International Geophysical Year for 1957 a small Earth satellite might be put up. This had just about become feasible. Towards the end of the Second World War Germany had developed the V2 (or A4 in Germany) rockets for long-range bombardment, and after the war the engineers and scientists concerned had largely migrated either to America or Russia. Indeed, the top American rocket expert, Dr Wernher von Braun, came from the German Peenemünde rocket establishment. The Americans had developed further these V2 rockets firing instrumented packages into quite high-altitude lobs, and sufficient know-how was at hand to develop a rocket suitable for putting up a small Earth satellite. The announcement of this intention was made on 29th July 1955, under the title of Project Vanguard.

Although the Americans had used the name Vanguard, it was the Russians who put up the first Earth-orbiting satellite in their famous Sputnik I on 4th October 1957, and the Americans were hard pressed to put up their first satellite four months later. The sizes of these satellites were small by today's standards–184 lb for the Russian and 31 lb for the American–but there was no doubt in the minds of most progressive and responsible persons that the space age had dawned. Suddenly, all the projects discussed in theory by the interplanetary society enthusiasts seemed capable of realisation–from the creation of satellite communications networks to men visiting the planets.

Quietly, behind the scenes, experts in America began assessing the possibilities. Of immediate interest were the military ballistic missile rockets Jupiter, Thor and Atlas, all of which could possibly be modified for use in a space role.

The United States decided, however, right at the start that space exploration should be a purely civil venture divorced from the military rocket programme and that US activities in space should be devoted to peaceful purposes for the benefit of mankind. Accordingly, in October 1958 the National Aeronautics and Space Administration, known the world over as NASA, was set up. This took over the National Advisory Committee for Aeronautics, and it inherited a number of rocket vehicles and space-flight programmes from the defence services, some aeronautical research projects and the scientific goals of space exploration which had emerged from the International Geophysical Year. NASA started immediately to build up an ambitious programme embracing the space sciences, planetary investigations, manned space flights and a number of projects with immediate practical use. This programme gained momentum, and in 1960 NASA outlined its ideas for manned exploration of the Moon and nearby planets.

Only one year later the decision was made and the United States Administration assigned to NASA the major role of landing men on the Moon and bringing them back safely by the end of the decade.

This decision was, of course, based on studies which had been going on for 2 years or so, studies which gave some indication of the kind and amount of work involved and some ideas of cost. A rocket vehicle three, four, five or even more times as powerful as the biggest then under development would be needed. Special manufacturing, assembling and launching facilities would need to be worked out and built; a vast network of tracking stations and communications centres would have to come into being; men would have to be trained and launched into Earth orbit and perhaps into Moon orbit as part of the development programme. There was so much to work out and so many factors interrelated that at the start a detailed plan could not be produced. As investigations and calculations proceeded, certain major decisions were able to be taken and the general plan began to take shape.

In 1960 the programme appeared thus: manned Earth orbital flights in 1961, landing small payloads on the Lunar surface by the Centaur rocket about 1961–2 and onwards, the introduction of the Saturn rocket with 1½ million lbf thrust in 1962–3, manned satellite development by 1963, Nova rocket with 6 million lbf thrust in 1965–8, orbiting space station by 1970, manned circumlunar flight by 1970 and a manned lunar landing shortly after 1970. By any reckoning, it was an ambitious programme and one which caught the imagination of many people.

Even at this early stage it was reckoned that perhaps the skill of 400,000 workers, the know-how and capability of some 20,000 firms and the brains and experimental facilities of 150 universities and research groups would be needed. Not least, management techniques had to be developed to control this vast and widespread effort. The total cost over 9–10 years was reckoned initially to be about $20,000 million or £8000 million.

It would, of course, have been unwise, to say the least, to embark on such a programme without positive evidence that man could exist reasonably comfortably out in space, at least for the lengths of time needed for the Moon mission. How would man react to weightlessness? For as everyone knows today, when a spacecraft is in the condition known as 'free fall' both it and its occupants are moving identically under the same forces so that there is no reaction between them and hence no 'weight'. Some persons at the time even thought that men might become mentally unbalanced shut up in a capsule out in space. Consequently much had to be found out and found out quickly, because it was no good going ahead to provide very expensive hardware and systems if man himself was not physically or psychologically up to making the Moon trip.

Consequently, some time before the great decision was made in 1961, a series of manned flights had been planned under the heading of Project Mercury–indeed, the beginnings of this project went back to the very first week of NASA's formation in 1958.

The prime objectives of Project Mercury were to place a manned spacecraft in orbital flight round the Earth, to develop a pool of trained manpower which could create the basic technology for manned space flight, to develop launch, control and recovery techniques, and to verify man's capability in the space environment.

The first American rocketed into space was Alan B. Shepard, who made what was called 'a sub-orbital lob' lasting about 15 minutes which landed him some 300 miles out in the Atlantic on 5th May 1961. The launch vehicle was a modified Redstone ballistic missile rocket, a direct descendant of the V2 rockets but much bigger. It was nearly 6 ft in diameter, 83 ft overall height, including the spacecraft, and had a lift-off weight of 66,000 lb, the single engine generating a thrust of 78,000 lb. For comparison the original V2 stood 46 ft high and weighed 28,500 lb at take-off.

Virgil I. Grissom followed with a similar flight, after which came four Earth-orbiting flights, for which modified Atlas ballistic missile rockets were used. The Atlas D used for these flights developed a lift-off thrust of 367,000 lb, more than five times that of the Redstone booster vehicles, and accelerated the Mercury capsules to about 17,500 miles/hour in their Earth orbits. The astronauts were John H. Glenn, Jr, M. Scott Carpenter, Walter M. Schirra and Leroy Gordon Cooper. The first two each did three Earth orbits, the third did six and Cooper stayed up for 34 hours and twenty-two orbits.

The success of these flights was very heartening. No really serious problems–engineering, biological or psychological–had shown up, and the pessimism expressed by some seemed to be unfounded, although many further experiments and tests were to be carried out.

Once again Russia stepped in with a 'first', Juri Gagarin being the first man to orbit the Earth on 6th August 1961, the American orbital flights starting later on 20th February 1962. However, the US programme continued with its series of ever more progressive shots, until by mid-May of that year all Project Mercury aims were considered accomplished, and plans for succeeding phases began maturing.

Project Mercury might seem a small event on the road to the Moon, but it was, of course, very important. Though it was one thing to place an instrumented package in orbit, it was quite another to send a man up. Man needs an environment not far removed from that to which he is accustomed on Earth if he is to survive, and placing men in space meant creating artificially a suitable mini-environment.

Designers did not have to start entirely from scratch. Some experience was available as a result of Major D. Simons' balloon flight over 19–20th August 1957, when he was sealed into a capsule for 32 hours and spent 5 hours at an altitude of 101,000 ft. As he said then, '. . . the conditions encountered were physically equivalent to those in a manned satellite in terms of the . . . ambient pressure, incoming radiation, heat balance and control, and in terms of the emotional reaction towards the overall flight plan'. The only really significant factor missing then was weightlessness; and, naturally, he had not been subjected to the acceleration forces of a rocketed astronaut.

There are two ways of creating a mini-environment for a spaceman–as indeed there are for deep-sea diving and similar activities. One can provide a man with a pressure

insulation layer was placed between the base of the spacecraft and the cabin itself so that in the short time of re-entry, the temperature rise in the cabin would remain within acceptable limits.

Ablative cooling is a sacrificial method, in that a part of the spacecraft is designed to be sacrificed in order to save the rest. The method worked and, as we shall see later, it has played an important part in making current space travel possible.

The final shape chosen for the Mercury spacecraft was thus rather like a bell, though it passed through many other shapes during its wind tunnel tests. It was a one-man vehicle some 9½ ft in height and 6 ft in diameter at the widest portion of the base where the re-entry shield was located. Its weight at lift-off was about 3000 lb.

It is worth noting the large amount of testing and development work carried out to make this project a success, particularly because many of the basic concepts created and tested here were carried over into the following generations of spacecraft.

A solid-propellent rocket, called 'Little Joe' was used for initial rocket-boosted flights so that the capsule and its components could be thoroughly checked out, and before any manned flights were undertaken three monkeys and an instrumented 'mechanical man' were sent aloft.

An escape system, consisting of a cluster of small rockets fixed to a tower on the capsule, was fitted so that in the event of an emergency the spacecraft could be fired clear of the main rocket. In normal flight, after a certain height was attained, this escape system was automatically jettisoned. This system was tested both on the ground and during booster-rocket flights. A parachute system had to be devised to bring the capsule softly down to earth once it had entered the denser layers of the atmosphere–essentially a drogue chute, a main chute and a reserve chute. Drop tests were carried out from aeroplanes to check on drogue operation, parachute deployment and so on, and small-scale models were projected from supersonic aircraft at great heights.

Also, it had been decided that it would be safer and more manageable for these spacecraft to come down in the sea rather than on land; consequently buoyancy equipment had to be provided, and dummy splash-downs and tests on flotation properties had to be carried through, while a whole series of rescue and recovery operations had to be worked out and checked in practice.

With the accent on using already available equipment wherever possible, designers started off with the hope of adapting aircraft equipment, but they soon found that in many cases this would just not do the job. Near absolute vacuum, weightlessness and extremes of temperature made equipment react differently from how it did in aircraft within the atmosphere, and equipment had to be tested in advance in a simulated space environment. These tests soon altered preconceived ideas on constructing–and testing–spacecraft. Although general aircraft design *philosophy* could be adapted, when it came to hardware, new designs to stricter specifications were often needed.

As the Mercury spacecraft was about the simplest one could imagine, integrating it with its launch vehicles–Redstone for the initial sub-orbital flights and Atlas for the orbital ones–was left until late in the programme, as this seemed a simple operation. In fact, this integration became quite troublesome introducing structural dynamics problems, arising partly from the effect of the protruding escape tower mounted on the spacecraft.

One of the most important aspects of Project Mercury was the complete change of man's role over the period of the flights. Initially, man's capability in space was an unknown factor. For this reason, and also because monkeys were to be sent up first, all the critical spacecraft functions were made automatic, so that the astronaut was thought of essentially as a passenger. However, on the first Mercury manned orbital flight, some of the thrusters which oriented the spacecraft became inoperative, and John Glenn assumed manual contol to complete the full planned three orbits. This was followed by the other astronauts taking over quite complicated functions when hardware malfunctioned, and it soon became clear that man was not only able to carry out useful work in space and make observations, but that he also provided an amount of redundancy–that is, 'back-up' for the various systems–which could not be achieved in other ways.

This brief summary only highlights a few of the bigger tests carried out during the development stages; in fact, every component and sub-assembly had to be tested in real or simulated conditions before acceptance. Costly and time-consuming as this development may seem–4 years of effort in which twenty-five flights were made, six of them manned, at a cost of nearly $400 million or £166 million–the information gained and the techniques worked out were to form the base for subsequent American space activities. At the same time the habit of exhaustively testing and developing hardware and techniques, and of practising operations, has stayed with NASA throughout its manned flight operations and has led to levels of reliability undreamed of even a few years ago.

The expenditure on Project Mercury worked out at about $100 million or £40 million for each of the four years, but this was simply a matter of 'getting their feet wet'–when the Americans took the Project Apollo 'plunge' it was estimated that the total cost of putting astronauts on the Moon would amount to $20,000 million or £8000 million. This is a massive figure by any standards and is near enough a quarter of the annual Gross National Product of the United Kingdom. However, with this expenditure spread over nearly 10 years, even at the peak period of work the cost would be around ¾% of the annual United States' Gross National Product, and at other times would be nearer ½%.

Why should it take 20,000 firms employing 400,000 persons and an expenditure at least ten times that of developing the Concorde supersonic airliner to put a few men on the Moon? This can only be understood as the story unfolds, and as the problems and their solutions are appreciated.

2
The Basic Problems

To appreciate the magnitude of the task which faced those engaged on turning the idea for Project Apollo into a reality, it is as well to see what the basic problem is in very general terms.

When a cricketer hits or throws a ball across the field, for a few brief seconds the ball is a ballistic missile – a body moving freely through space under the influence of the Earth's gravity. We may not usually think of it in this way but, as Newton pointed out, there is nothing essentially different about an apple falling from a tree, a ball thrown into the air, or the Moon orbiting the Earth – or, indeed, a spaceship travelling to the Moon. All are bodies moving freely through space obeying the law of gravity; the main differences are those of size and energy.

A thrown cricket ball, weighing $5\frac{1}{2}$ oz might reach a height of 100 ft, the total energy imparted to it being around 35 ft lbf. If a cricket ball were fired from a gun like a shell it might have an energy of 41,000 ft lbf and rise some miles before falling back to Earth. But how much energy would be required to throw the cricket ball on to the Moon? The answer is about 7 million ft lbf, and this is equivalent to giving it a speed of 40,000 ft/sec or 25,000 miles/hour. This is the critical 'escape velocity' below which any object leaving the Earth will fall back.

The only way to give a body of any size this kind of velocity is to accelerate it by a rocket. Some early science-fiction writers postulated firing a spaceship to the Moon from a gun, but they had obviously not done their sums, because even with a gun barrel 1000 ft long, any occupants of the spaceship would be completely flattened by the brief acceleration of around 25,000 *g*!

To propel men into space one has to consider accelerations of not more than, say, 4–8 *g*, depending upon how long the accelerating force lasts. A spaceship sent up vertically with a steady 4-g acceleration would attain escape velocity after about 5 minutes when it was some 2000 miles above the Earth's surface. Since for most of this time the body would be well outside the atmosphere, there really is no alternative to a rocket vehicle at the present state of knowledge.

A rocket propels itself in one direction by throwing out some of its mass in the opposite direction, relying on the physical law that every action has an equal and opposite reaction. In practical terms the part of the rocket thrown out is the fuel it carries, and to gain the greatest momentum from it, this fuel is burnt with an oxidiser in a combustion chamber to form very hot gases at high pressure, which escape through a specially-shaped nozzle at supersonic speed. The reaction to this jet is called the thrust.

At the time of lift-off, a very large proportion of the thrust generated is needed to balance the weight of the rocket vehicle–that is, simply to neutralise the effect of the Earth's gravity–and only the residue is available for accelerating the vehicle. As the rocket rises, however, fuel is used up and the mass being moved becomes less, so that the same thrust gives an increasing acceleration, which may start at $\frac{1}{4}$ g and rise to 4 g or more.

Rather than blast off straight for the Moon, it is more practicable to carry out the journey in clear stages after each of which the situation can be assessed before embarking on the next. This is done initially by arranging to put the spacecraft into an Earth 'parking' orbit about 100 miles up where, travelling at about 17,500 miles/hour in a circular path, its centrifugal force just balances the terrestrial gravitational pull. If at the time of orbital insertion the speed is in excess of that required, there is an imbalance between the centrifugal and gravitational force. The resulting orbit is then elliptical, the spacecraft alternately climbing away from the Earth and then falling back. The lowest point of the orbit is called the perigee and the highest is the apogee. As the excess speed at perigee increases over that needed for a circular orbit, so the apogee of the resulting elliptical orbit moves farther out into space, until when the 25,200 miles/hour escape velocity is reached the apogee is at infinity and the path has become parabolic.

After the astronauts have spent, say, two revolutions in Earth orbit checking their spacecraft, they can then move on to the next stage by accelerating up to escape velocity. The

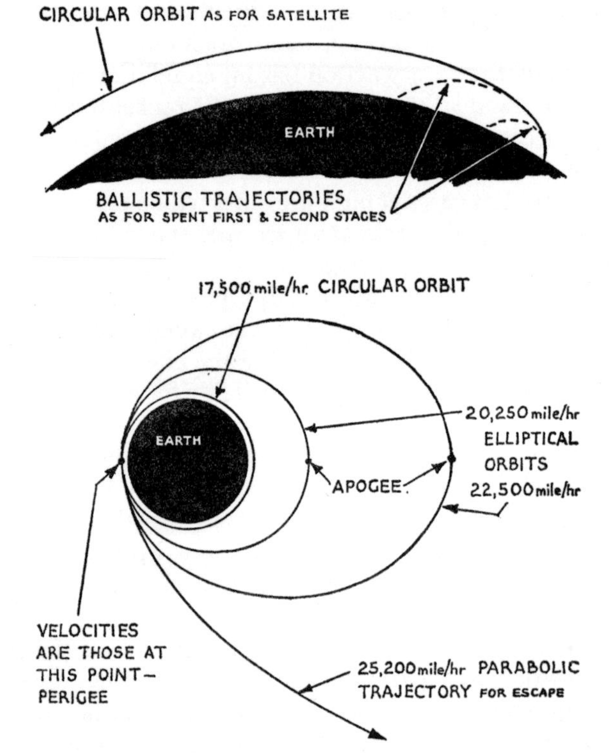

Fig. 2. Orbits and Trajectories.

spacecraft is then, as they say, injected into a trans-lunar trajectory, which is just like the path of the cricket ball across the field, but on a much grander scale.

The craft then coasts freely for most of the journey. As it climbs out into space against the Earth's gravity it slows down but, being aimed at the Moon–or more strictly speaking towards a position in space at which the Moon will be when the spacecraft gets there – a point is reached when the gravitational pulls of the Earth and Moon are equal. Beyond this point only the lunar gravity is significant and the spacecraft falls towards the Moon gaining speed as it falls. Again, instead of making a direct landing, it is more expedient to reduce velocity at the appropriate time and go into an orbit round the Moon. This is another 'parking orbit' in which the crew can take stock of the situation. The next phase would be reducing speed still further so that the spacecraft falls towards the lunar surface. As the Moon has no atmosphere, neither parachutes nor wings can be used for landing, and the craft would have to slow itself down by directing its rocket thrust downwards for a soft landing.

The escape velocity of the Moon is much less than that of the Earth – only about 5100 miles/hour – so that the energy requirements are very much less for landing and taking off than they would be on Earth. The return trip to Earth would go through similar stages, and there would probably have to be mid-course corrections in both coming and going to make up for any inaccuracies in the original aimings.

All round, it can be seen that although the bulk of the trip there and back is 'coasting' simply as a missile on a trajectory in a gravitational field, rocket power has to be used a number of times to get up into and out of parking orbits and trans-lunar and trans-earth trajectories. The amount of fuel burnt depends upon the change of velocity necessary, and on the size of the mass being accelerated, as well as on the efficiency of the rocket engine. The mass accelerated in the early phases consists not only of the eventual payload and the rocket structure but also all the fuel needed for later manoeuvres. Initially, therefore, the mass to be lifted off the Earth's surface can be quite large for even a moderate mission, calling for a very large launch vehicle.

It would be ideal really if the actual rocket structure could be consumed at the same time as the fuel, so that there were no empty tanks left as 'dead weight'. An approximation to this ideal is attained in practice by building rocket vehicles in stages so that each stage is jettisoned as soon as its fuel is exhausted and its work is done. In theory one could have any amount of stages and attain a high efficiency, but in practice this number is limited because of the complexity involved and the fact that each stage has to have its own rocket engines. Three-stage vehicles have become most familiar among the Earth satellite launchers, though some have been only two-stagers, and up to five stages have been proposed for very big rockets.

The formalised rockets in Fig. 3 will probably help to put this matter of fuel and payload into perspective. In each case the rocket is considered idealised into a cone. Case (A) is that of putting a payload into low Earth orbit. The black top portion represents the payload and the shaded portion the fuel and structure necessary. Case (B) is that of attaining escape velocity–i.e. of transferring from the Earth orbit to a Moon trajectory. (This is less than the escape velocity needed to reach the planets.) The payload of (A) now has to be considered as split into two portions–a new smaller payload and the fuel needed to acquire this extra velocity. Case (C) is that for soft landing on the Moon. The payload of (B) is again split up into a new yet smaller payload and the fuel required for braking the vehicle. Case (D) is that for returning from the Moon. The mass landed on the Moon in (C) now has to be considered as the return spaceship and the fuel needed to blast off the Moon. Finally, in (E) we have the case many feared would be necessary even twenty years ago or less–that of returning to the Earth and needing rocket power to effect a landing, it being assumed that the spacecraft would otherwise burn up in the atmosphere like a meteor. The mass now blasted off from the Moon consists of a very small payload indeed and a very large amount of fuel, the latter being required to slow down the spacecraft from its return speed of about 25,000 miles/hour to something in the region of hundreds of miles/hour so that parachutes can be used in the final landing phases.

One can see in these five cases how the actual payload is progressively reduced. To get this even further into perspective, these launch vehicles are redrawn at the bottom of Fig. 3 so that the payloads carried are to the same scale. One

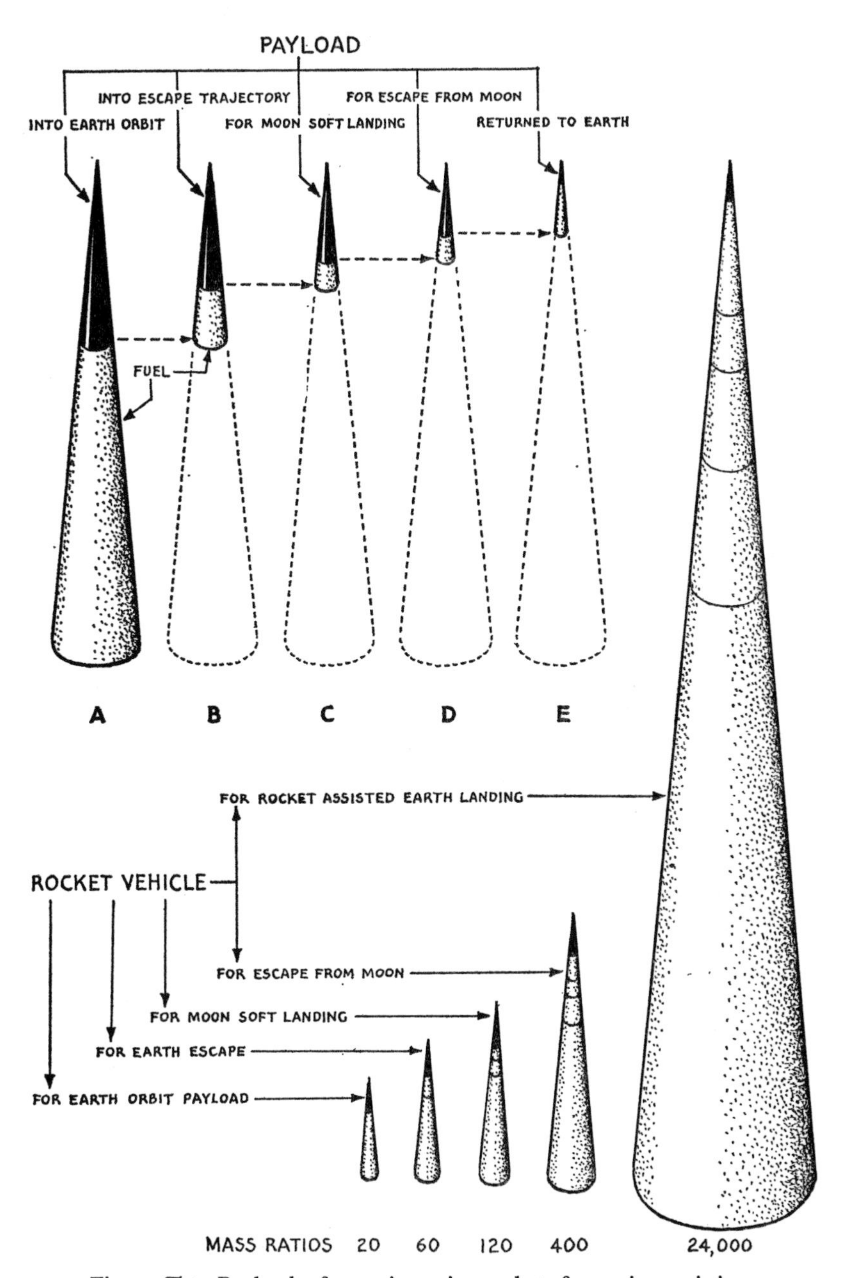

Fig. 3. *Top:* Payloads, for a given size rocket, for various missions. *Below:* Sizes of rockets for the various missions carrying a standard payload.

notices immediately that case (E) is a sheer impossibility. These idealised rockets have been calculated on the basis of a structural efficiency similar to the best contemporary models, and the ratios of take-off mass to payload, called the mass ratios, are respectively 20, 60, 120, 400 and 24,000. In the last case it would take a rocket of 24,000 tons just to carry a load of 1 ton to the Moon and back; and to transport a worthwhile spaceship of, say, 10 tons would require a rocket of some quarter of a million tons!

The method of using an ablative shield to protect a spacecraft entering the Earth's atmosphere direct from space has luckily made case (E) unnecessary, and case (D) is a much more practicable proposition.

In the 1930s the situation did not look so promising. There being no rockets to mention, structural efficiency had to be 'guestimated' on the basis of the current aircraft structures. Fuels were also then a problem and the calculation of take-off weights were considerably greater than those suggested above. Ingenious minds naturally sought ways out of the dilemma. One study assumed that the returning spacecraft had short wings and could use these for aerodynamic braking in the Earth's atmosphere on return from space and for gliding down to the Earth's surface–and we may yet see something of this kind materialising.

At that time Ing. Guido von Pirquet suggested another approach. Firstly, a number of practicably sized rockets would be used to ferry fuel and supplies out to a 'space station' in orbit round the Earth; secondly, the space rocket proper of reasonable size, having only to lift itself into Earth orbit, would come alongside the 'space station' for refuelling and provisioning. When it started on its trip to the Moon it would already be travelling at over 17,000 miles/hour and, for a 5-ton payload, its mass could come down to something like 1000 tons. Many variations of this method were, of course, suggested. The simplest replaced the 'space station' and supply rockets with just one large supply rocket which went up into orbit and waited for a second rocket with the spaceship to join up with it for fuel transfer. All of these ideas were eventually classed under the heading of Earth Orbital

Rendezvous techniques, because they depended upon a meeting in Earth orbit.

From the 1930s right on up to the beginnings of the American space programme there had been two accepted methods of getting a spacecraft to the Moon–remembering that the 'spacecraft' in this context includes the rocket to bring it back again. These were the Direct method using one booster rocket, and the Earth Orbital Rendezvous method using two or more rockets. However, before a choice of methods could be made, the very tractability of the problems led to yet a third method being invented, as we shall see.

3

The Plan Takes Shape

IN 1961, when the decision was made to go ahead with Project Apollo, there was only one large rocket–known later as Saturn I–being developed in America. This had been conceived in general terms by Dr Wernher von Braun and his team at the Army Ballistic Missile Agency establishment at Huntsville, and development was approved by the Department of Defense's Advanced Research Project Agency. The Huntsville establishment was taken over by NASA in July 1960 and its title was changed to the George C. Marshall Space Flight Center, Dr von Braun being made Director. In 1961 the Saturn I's role as a test vehicle for the Apollo programme was also defined.

The development of the Saturn I was in stages. Briefly, the original version, called Block 1, was to have dummy upper stages and be used just to prove the basic concept. The first rocket stage consisted of eight up-rated engines from a Jupiter-type military rocket vehicle with eight Redstone rocket-type tanks of 70 in diameter clustered round a single 105 in Jupiter tank for fuel and oxidiser. The Block 2 version was to have up-rated H-1 engines in the first stage giving about 1½ million lb of thrust, a live second stage designated S-IV and a dummy Apollo spacecraft. This was to be followed by the Saturn IB, which had a Chrysler-built first stage with engines uprated even further, and a new second stage designated S-IVB. This would stand 224 ft high compared with the Saturn I's 190 ft, and would be about 650 tons fuelled up. The Saturn IB would be used for putting manned and unmanned Apollo spacecraft into Earth orbit for a series of tests, as well as for launching other equipment, some of which would only be indirectly connected with the manned space programme.

At a very early stage it was recognised that even this series of rockets would be woefully inadequate for the purpose of putting a man on the Moon, so NASA put in hand with the Rocketdyne Division of North American Aviation Inc. the task of developing a new rocket engine, designated the F-1, with a nominal thrust of 1½ million lb–as much as that pro-

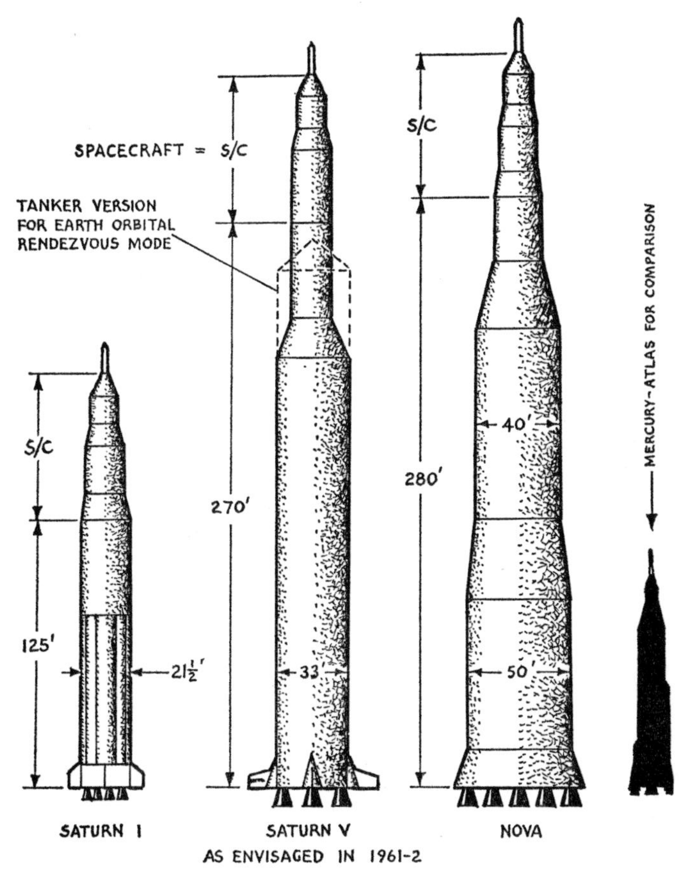

Fig. 4. General form of Saturn and Nova rockets as projected in 1961 compared with Mercury/Atlas rocket.

duced by the cluster of eight in the Saturn I's first stage. Whatever kind of rocket vehicle eventually emerged, these new engines would be able to be used in clusters to give what ever lift-off thrust was required.

Two modes, as they were called, were known for getting to

the Moon–the Direct Mode and the Earth Orbital Rendezvous Mode–and both were studied in depth. The Direct Mode did not get very far. By assuming a very advanced kind of structure and the best fuels that might become available, a vehicle of some 5000–6000 tons weight might be able to send up a 15-ton spacecraft to the Moon and back by the Direct Mode. This vast vehicle was given the name Nova, but it was never very much more than an idea on paper, though protagonists produced drawings of what it was expected to look like and gave very optimistic forecasts of its performance. In 1962 the conception for Nova was not unlike that of the Saturn family and it was intended that eight F-1 engines should be used for the first stage, giving a lift-off thrust of some 12 million lb. Nova would have been able to lift 200 tons into low Earth orbit or 75 tons into deep space. However, a rocket vehicle like this–ten times as big as the largest then under development and forty times as big as the Atlas, which was used to launch the orbiting Mercury spacecraft–was a lot to ask for in a field where progress had so far necessarily been through fairly conservative steps. While it would no doubt have been possible to produce such a rocket vehicle, when it became apparent that a vehicle of this size could not be developed within the timescale set for Project Apollo, it had to go at the bottom of the list, and with it the Direct Mode.

While this mode was being investigated, von Braun's team were producing initial plans for the biggest rocket vehicle which they could conceive being developed within the allotted time. This was about four and a half times as large as Saturn I, and was designated Saturn V (originally Saturn C-5). The idea was to use as many common, and thus tested, parts as possible from the Saturn I and IB family, particularly with respect to instrumentation, electronics and so on. This new vehicle would have a much larger first stage, designated S-IC and using five F-1 engines, and a new second stage, the S-II, but would use a version of the S-IVB stage from the earlier Saturn vehicle. The Saturn V was to be a more slender vehicle than the ideas for Nova, and would weigh between 2700 and 3000 tons at take-off. NASA approved the development programme for the Saturn V on 25th January 1962.

In examining the Earth Orbital Rendezvous Mode, techniques were studied which would exploit either or both of the Saturn vehicles, because it would have been patently undesirable to embark on yet another. The simplest method worked out required two Saturn V vehicles. One would carry the Apollo spacecraft into a parking orbit round the Earth, and would have a partially fuelled third stage. A second Saturn V would then go up to rendezvous with the first, the third stage of this being used as a liquid oxygen tanker for topping up the vehicle in orbit, which would then be capable of making the voyage to the Moon and back.

There were some advantages to this scheme; no huge carriers would be needed and, although the refuelling in space might prove to be a difficult exercise through weightlessness, the experience could be applied to later missions for journeys to the planets and so on. On the other hand, without experience there was no saying how difficult it might prove to match up two very massive objects in Earth orbit. The method needed two Saturn V vehicle launches within a short time of one another. What if the second launch failed in some way? How long would it take to get another one ready? What would happen to the one already in orbit in the meantime? Although in 1962 this seemed the most likely method, it was questions like these which caused NASA engineers not to rush at this approach. They would probably have had to accept it, however, if the reluctance had not 'squeezed out' a third mode.

The new idea came from John C. Houbolt, of the Langley Research Center, early in 1962. An examination of the whole series of events from Earth blast-off to a safe return landing showed that there was a considerable loss in overall efficiency resulting from landing on the Moon, and blasting off again, a lot of mass which was really redundant for a simple lunar exploration. For instance, a complicated parachute system would be needed for landing on Earth. This had to be carried all the way to the Moon and then back just for use in the last quarter of an hour. Likewise there were navigation and computation equipment, fuel and environmental supplies which would be needed for the return trip but which were not actually needed on the Moon's surface. So, asked Houbolt, was it really necessary to use a considerable quantity of fuel to land this equipment on the Moon and then boost it off again? Would it not be possible to divide the spacecraft into two parts, with the greater part remaining in orbit about the Moon, thus retaining a velocity of about 3600 miles/hour, and only land the part necessary?

Because the part landed on the Moon would, after blast-off, have to rendezvous with the part still in Moon orbit, this idea and variations on it became known as Lunar Orbital Rendezvous Modes. There are, of course, a lot of different ways of splitting up a spacecraft. At one extreme, one could leave in orbit nothing more than a tank of fuel for the return trip, the spacecraft proper carrying out the lunar landing and rendezvousing with the tank for fuel transfer prior to Trans-Earth Injection, as changing from a lunar orbit to an Earth-bound course became known. At the other extreme, one could retain as much as possible in orbit, landing the bare essentials in a minimum craft. One of the initial concepts for the lunar landing part was, in fact, a vehicle of 1800 lb with an astronaut using 'a plumb bob, a stop watch and a reticle' as one NASA official described it. A more realistic approach stipulated a landing module of about 24,000 lb–though events, in time, showed this to be rather conservative.

This third mode had to be examined in great detail, especially in comparison with the Earth Orbital Rendezvous Mode. All told, in the examination of all three modes, over a hundred engineers spent about a million computer-aided hours in study before recommending the Lunar Orbital Rendezvous Mode as the best in terms of the greatest safety and reliability, the least expense and an assured development time within the period of NASA's target date, the end of the decade. The decision to proceed by way of this third mode was made in July 1962. Although there was a long way to go, at least the route was now known.

In the meantime, work had been going on with respect to the follow-on from Project Mercury–a two-man spacecraft conceived initially before the first Mercury orbital shot and given the name Gemini, the twins. Its initial aims were to expand the technology of manned space flight and to demonstrate long duration space flight, as was consistent with the overall aim of building up a viable space capability. Added to this was the carrying out of experiments in rendezvous techniques, and this became especially important with the acceptance of the Lunar Orbital Rendezvous Mode, when the aims were extended to cover the actual 'docking' of one craft with another.

This is a more complex business than might be thought at first sight. Suppose a spacecraft and a target vehicle are in the same circular orbit about the Earth but separated by, say, 200 miles. With an aircraft operating in the atmosphere the chaser would catch up by increasing its speed. If the spacecraft makes a substantial increase in speed, instead of catching up the target, it will go into an elliptical orbit rising above the target's orbit. One would conclude that it would now pass over the top of the target. In fact, it would never catch up the target, for its new elliptical orbit would take longer to traverse than the target's circular one, and when the spacecraft returned to its perigee (lowest) position, it would be even farther behind the target vehicle. This, naturally, suggests a method of catching up–slow down so that one changes into an elliptical orbit below the target, which will shorten the orbit time. If the distance is, as we said, only 200 miles, at an orbital speed of about 18,000 miles/hour, the spacecraft only has to pick up 40 seconds in one orbit of $1\frac{1}{2}$ hours.

The procedure, of course, becomes more difficult if the two bodies are in orbits not in the same plane, or at different heights, and a computer becomes necessary, especially if it is required to work out an optimum transfer mode which will reduce the amount of fuel expended–it might well prove to be more economic to catch up slowly spread over a number of orbits than make one jump at it.

This is not a matter which need be followed up here in detail, but the complexity of matching up in orbit comes out well by noting the manoeuvres planned for a Gemini spacecraft attempting to rendezvous with an Agena rocket-target vehicle–an exercise carried out in this Gemini programme.

The Agena is in a circular orbit 185 miles up. The Gemini spacecraft is in an elliptical orbit with a perigee of 100 miles and an apogee of 168 miles. The catch-up rate is 6·68 degrees/orbit–1·67 minutes of time or 468 miles. When the distance is closed sufficiently the Gemini reduces its catch-up rate to 4·51 degrees/orbit by increasing its speed at apogee so that while its highest point is still 168 miles, its lowest is raised to 134 miles.

Near the third orbit apogee (highest point), the Gemini crew circularise their orbit at 168 miles, their spacecraft being only 161 miles behind the target vehicle, though their orbit is 17 miles below the Agena's. Target is locked-on by radar and the crew switch computer to rendezvous mode and begin terminal phase systems check-out. When range is reduced to 39 miles, the spacecraft is tilted up about 27 degrees and thrusters are ignited to propel Gemini towards Agena target vehicle. Twelve minutes later the computer displays the first correction to be applied. After a further 12 minutes another correction is made. With the range down to 4½ miles, the crew begin a semi-optical approach, and at a range of 2000 ft, the command pilot reduces the closing speed to 4 ft/sec (about 3 miles/hour), finally reducing this to less than 1 ft/sec until the craft meet. The time for the complete operation is about 5½–6 hours.

It will be appreciated that a great deal of manoeuvreability was to be demanded of the Gemini spacecraft, especially for the final docking operation, when it would actually mate with the target craft. The spacecraft would need fine control in roll, pitch and yaw, and be able to thrust itself sideways, up or down. The kind of speed changes necessary for changing orbits – about 50 ft/sec maximum – would be small compared to its orbiting speed – about 25,000 ft/sec – but would nevertheless need quite powerful thrusters, and something like 600 lb of fuel and oxidiser would be needed just for this manoeuvring.

These factors, of course, had their effect on the Gemini

spacecraft design, as did the launch vehicle chosen. This was a modified US Air Force Titan II Intercontinental Ballistic Missile, which had two stages 63 and 27 ft in length and both 10 ft in diameter, the first stage with two engines having a lift-off thrust of 430,000 lbf. This was chosen partly because it used storable hypergolic propellent. The fuel and oxidiser burn on contact so that no ignition is necessary, and if there is a hold-up in launching the fuel does not evaporate, need topping up or unloading. These properties made it ideal for lifting off very exactly at predetermined times to effect orbital rendezvous.

The Gemini crew's cabin was similar in shape to the Mercury craft, though bigger – 11 ft high on a $7\frac{1}{2}$-ft base compared with $9\frac{1}{2}$ ft on a $6\frac{1}{4}$-ft base. However, this was not the whole spacecraft. In this case another section was added to contain equipment, so that the craft at launch consisted of a Re-entry Module and the Adapter Section – the latter having this name because its shape was such as to adapt the spacecraft proper to the Titan vehicle.

The Re-entry Module itself consisted of three parts. At the top, or front as it would be in orbit, was the Rendezvous and Recovery Section, which contained the drogue, pilot and main parachutes for landing, and radar. Next to this was the Re-entry Control Section, which contained fuel and oxidiser tanks, valves, tubing and two rings of eight attitude control thrusters each for control during the re-entry phase. The main parachute was attached to an adapter assembly in this section. Behind this was the Cabin Section, which housed the crew of two side-by-side, their instruments and controls. The dish-shaped ablative shield formed the large end of this section.

Behind this was the Adapter Section, $7\frac{1}{2}$ ft high and 10 ft diameter at its base, and itself made up of two sections. The Retrograde Section contained four solid propellent retrograde rockets, which were used for slowing the spacecraft down for re-entry, and part of the radiator for the cooling system. The Equipment Section behind this contained batteries for electrical power, fuel for the orbit attitude and manoeuvre system, primary oxygen for the environmental control system, and a cooling system, the section itself acting as a radiator for this system. This Equipment Section was jettisoned immediately before the retro rockets were fired for re-entry, and the Retrograde Section was jettisoned immediately afterwards.

In the Mercury craft, through space and weight limita-

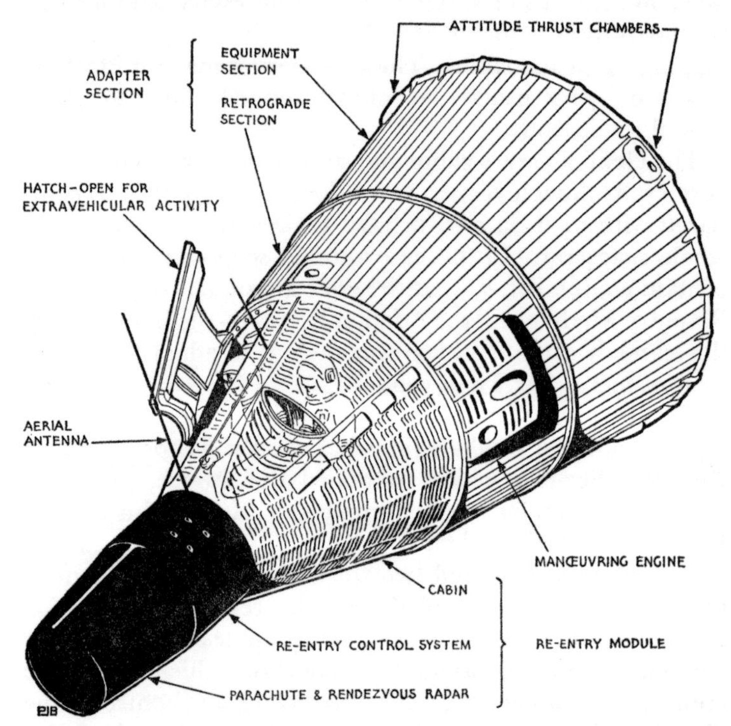

Fig. 5. General view of Gemini Spacecraft.

tions equipment was packed on top of other equipment in layer cake style. This caused great difficulties in maintenance as, to get at one system, many others might have to be disturbed and so have to be rechecked afterwards. In the Gemini design every attempt was made to segregate systems and equipment so that there was the minimum of interference.

The fuel and the attitude thrusters in the Re-entry Control

Section were for orienting the spacecraft so that it could be manoeuvred into the correct attitude for re-entry into the Earth's atmosphere, i.e. with the ablative shield forward.

The fuel and thrusters in the Equipment Section were for control of the Gemini spacecraft in normal flight. These thrusters, seen in Fig. 6 were really small liquid-fuelled rockets; they were of two kinds, giving 100 lbf thrust for

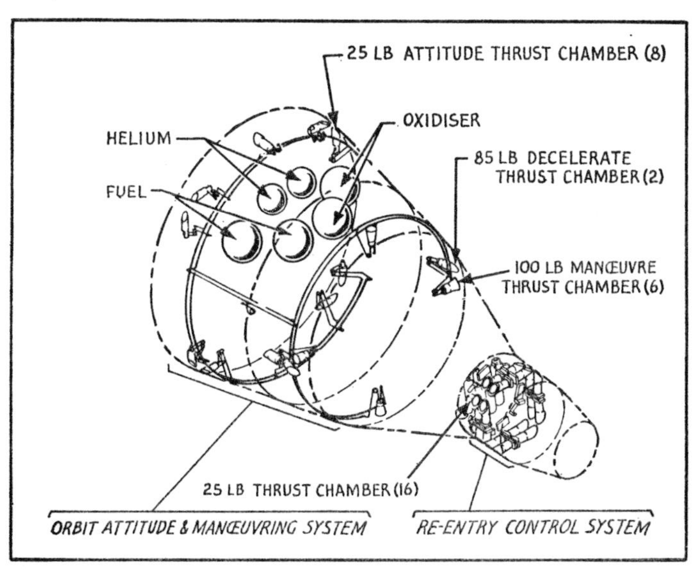

Fig. 6. Positions of Gemini Spacecraft thrusters. The Re-entry Control System is used when the other modules have been jettisoned.

manoeuvre control and 25 lbf thrust for attitude control. The rear 100-lbf thrusters were used to give the velocity changes necessary in transferring from one orbit to another.

The crew's Cabin Section consisted of a titanium pressurised hull fitted inside an outer beryllium shell, which was corrugated and shingled to provide aerodynamic and heat protection properties, a feature used on the later man-carrying Mercury capsules. Equipment not requiring a pressurised environment was located between the hull and the outer shell.

While the early models used batteries, positioned in the Re-entry Section and the Equipment Section, later models used fuel cells. Here energy was produced by forcing the reactants into the fuel cell stacks where they were chemically changed by an electrolyte of polymer plastic and a catalyst of platinum. The resulting ions combined with oxygen to produce electricity, heat and water. These fuel cells were rather experimental and were backed up by batteries, but their advantages for space flights were very apparent–a source of electrical power whose by-products, heat and water, were of value to the crew and systems.

With the Gemini spacecraft an opportunity arose to try out different crew escape procedures in case of emergency. As in the Mercury Programme the bulky escape tower on top had created dynamic problems with respect to the rocket structure, this was done away with and aircraft-type ejector seats were fitted, one for each astronaut. At low altitudes, in the case of an emergency, each astronaut would open his own hatch, eject and come down by parachute. Beyond a certain height the whole Re-entry Module would come down by its own parachute. This new approach was possible partly because of the hypergolic fuel used in the Titan rocket. In the event of an accident to the launch vehicle, this fuel would not explode violently as was the case with the fuel used in the Atlas Mercury vehicles, where a method of taking the astronauts well out of the danger area quickly was imperative.

The McDonnell Aircraft Corporation, which had dealt with the Project Mercury work, became prime contractors for the Gemini craft. Twelve Gemini flights were scheduled and, although there were occasional malfunctions here and there, by and large they were amazingly successful. There were two unmanned Gemini flights on 8th April 1964 and 19th January 1965 to test the Gemini launch vehicle and the ability of the spacecraft to stand up to the launch environment and the re-entry heat conditions, and so on. The first manned flight was on 23rd March 1965 and the last on 11th November 1966. Gemini 4, the second manned flight, stayed up for sixty-two revolutions and was famous at the time when astronaut Edward H. White carried out what became known as Extravehicular Activity to the experts and 'walking in space' to the layman, using a handheld manoeuvring gun for the first time. This used freon gas as a reaction jet. Gemini 5 stayed up in orbit for 7 days and 23 hours.

Gemini 6 was to have been an exercise in docking with an Agena target vehicle, but it was held back from launch as the target vehicle had to be destroyed after launch due to a malfunction. As it was, Gemini 7 was next to go up on 4th December 1965 with Frank Borman and James A. Lovell aboard. The same launch pad was quickly made ready and Gemini 6 was launched on 15th December with Walter M. Schirra and Thomas P. Stafford aboard. Gemini 6 made a rendezvous with Gemini 7, the two craft approaching to with a foot of one another. They carried out a series of impressive space manoeuvres and took the first colour photographs showing spacecraft orbiting the Earth.

The first space-docking operation was carried out successfully on Gemini 8's mission, but soon after coupling up with the Agena target vehicle, pitching and rolling motions commenced and these persisted even after the craft separated, so that the mission was prematurely terminated. New rendezvous techniques were tested out by Gemini 9 and a fully successful docking took place with Gemini 10. Geminis 11 and 12 carried out between them six more dockings after rendezvous, and there were eight periods of extravehicular activity in the last four flights.

The programme was not, however, carried through without troubles and stresses. There were inevitably development difficulties and the work became ever more complex and expensive. Even at this stage there was opposition from those who objected to the ever-increasing cost.

Nevertheless, when the series had been completed, everyone was extremely happy with the progress made. A lot had been learnt in the technical and engineering sense, as well as in the spheres of biology, physiology and psychology of man in space, and many of the fears had proved to be unfounded in practice. Man could operate successfully in space, and a remarkable number of 'firsts' had been chalked up by the USA.

4

Spacecraft to the Moon

BEFORE committing a manned lunar craft to a lunar mission, it was essential to gather as much information as possible about conditions on the Moon. This was necessary in order to plan on one hand the details of the astronauts' final mission and, on the other, to ensure indeed that no characteristics existed on the lunar surface which were fundamentally dangerous to the astronauts, beyond those that were already known. NASA had indeed started on a programme at the beginning of the decade for examining the Moon through unmanned space probes and, after the decision to land men on the Moon, this programme was used increasingly towards gathering information for Project Apollo. These space probes belonged to the Ranger, Surveyor and Orbiter series.

The Ranger project consisted of a series of small spacecraft developed from the then Earth satellites. They were equipped with television cameras and contained a small survival pack with instruments for measuring various surface conditions on the Moon. The object of Ranger was to descend to the lunar surface during a reduced velocity descent phase in which the television cameras would observe the surface and transmit the information back to Earth before their destruction at impact, and also to deliver the survival package onto the Moon's surface. Through lack of experience, the first Rangers launched gave a lot of trouble and it was Ranger 7 which was the first to achieve a successful impact after producing the first close-up pictures of the lunar surface during the course of its final descent phase. This first successful shot will probably be remembered for the drama during the time of its direct television transmission in the Jet Propulsion Laboratory Observatory in California, where the responsibility for this programme had been centred. Few who saw it will forget the dramatic sight of the pictures growing in detail as the spacecraft hurtled towards the Moon at several thousand miles an hour.

After a few more successful Ranger shots came the more sophisticated Surveyor programme. This was aimed at soft-landing spacecraft on the Moon in order to acquire even

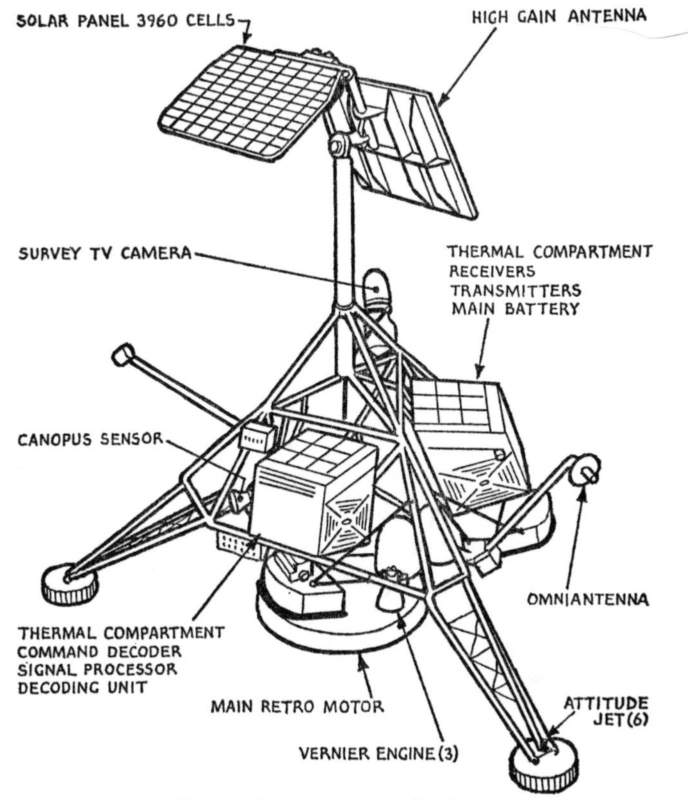

Fig. 7. Surveyor Lunar Probe.

more detailed and precise information concerning the surface conditions, particularly its mechanical properties, such as strength to support a spacecraft, etc.

The Surveyor craft were equipped with three extendable feet and scanning television cameras, the direction of which could be remotely controlled from Earth. The first Surveyor landed on the Moon after having been launched by an Atlas/

Centaur rocket from Earth on 30th May 1966–two years after the first successful Ranger craft–and it returned 11,000 fine photographs by television techniques. This was another historic 'first', the provision of detailed pictures of the Moon's surface taken actually on that surface.

At this time it was known that the Moon had no magnetic field and, of course, no atmosphere. It now seemed that the surface was covered in a powdery pumice-like dust with a few rocks of various sizes showing through. It was now evident that the surface below this thin dust covering could withstand the landing impact and weight of the Surveyor craft. This was welcome news because some pessimists had suggested that the dust might be very thick and spacecraft might just sink below the surface. Surveyor 2 was not successful, but Surveyors 3, 5 and 6 landed successfully on the Moon; in fact, Surveyor 6 (launched 7th November 1967) was subsequently lifted off and moved by remote control some 15 ft sideways by its rocket motors in an attempt to determine what effect the jet from the rocket had on the Moon's surface. One of the great unknowns had been the effect of the jet of the descent engine playing on the lunar surface, and to what extent disturbed dust might obscure the vision of the astronauts when they eventually landed, and whether it would damage the surface significantly as a landing base for a spacecraft. A lot of work was done on dust ballistics to determine exactly where the dust would go during a landing. In the case of the Surveyors themselves, being powered by solar cell arrays, it was important that dust did not land on these cells to impair their operation. None of these things in fact happened, and from the samples taken up by a small mechanical scraper by the Surveyors, photographed and analysed, it seemed reasonable to suppose that the Lunar Module could be safely landed on the Moon's surface.

However, more information was still required concerning possible landing sites, in addition to the detailed information derived from the Ranger and Surveyor projects. For this purpose the Lunar Orbiter project was developed by the Boeing Company, the aim here being to insert small unmanned spacecraft into lunar orbits, typically with a perigee of about 100 miles and an apogee of about 1000 miles. Each craft was equipped with several television cameras and was effectively a reconnaisance satellite designed specifically to cover the Moon's equatorial regions where five possible landing sites had been selected from Earth. The Lunar Orbiter craft were launched concurrently with the Surveyor programme over the years 1966–7 and they sent back to Earth some remarkably detailed pictures of general surface conditions prevailing in the crater areas and the maria, including those on the rear side of the Moon, which can never be seen from Earth. As a result of this very successful programme the five projected landing sites were confirmed and identified in more detail.

The Lunar Orbiters had a precise stabilisation system to enable them to effect the direction pointing of the cameras, and they carried a high-powered data transmission system to send the television pictures back to Earth.

The scene, then, was now set. Information from these unmanned spacecraft had confirmed, in general, the predictions of the scientists and engineers back here on Earth, and no new information had been brought to light which might alter the main course of planning for the Project Apollo mission. In particular, and most important, was the confirmation that, in general, the lunar surface was suitable for a spacecraft landing. The Americans who were planning the programme could not have hoped for a better set of data from these three types of unmanned probes.

We come now, therefore, to the design concept of the spacecraft for the main lunar missions–Project Apollo. From the basic ideas outlined earlier, it can be seen that in order to minimise the size of the initial launching vehicle, the project had to be conceived with the aim of returning to Earth the smallest-sized capsule consistent with the basic mission requirements. By definition this, clearly, should contain just the crew and the equipment necessary to effect their safe and accurate re-entry into the Earth's atmosphere and the subsequent descent to the Earth's surface.

NASA, in conjunction with various contractors, had for many years been examining the question of the optimum size of spacecraft and the number of crew which would be needed to man it, and the conclusion had been reached that a crew of three would be most appropriate. This decision was reached on the basis of the duties and tasks which would need to be carried out during the course of a lift-off, the flight to the Moon, the work on the Moon itself and the return journey to Earth. The first few journeys to the Moon would involve probably the minimum number of experiments, but account had also to be taken of the increasingly complex work schedule that would need to be accommodated on later flights.

In fact, the crew number of three was chosen before final details were worked out on the launch vehicle and the techniques that would be used to effect the actual landing, and it was subsequent to the choice of three as an optimum that decisions were finally taken which would involve only two of the three astronauts landing on the Moon.

The Apollo spacecraft, like Gemini, was devised on the modular principle. The first two parts are the Command Module and the Service Module. The Command Module is the part which houses the crew members and it was designed using the same general philosophy as was used on the successful Mercury and Gemini flights–namely, a truncated cone body with a curved ablative heat shield at the base and provision for an ultimate parachute descent in the lower part of the Earth's atmosphere. This Command Module and the Service Module can be seen in Fig. 8. The Service Module, as its name implies, provides the main services throughout the flight, including the basic propulsion system and the main part of the life support systems and power systems for the journey to the Moon and back.

In the original concept, the third module was to have been another rocket stage, which would have been used for landing the previous two modules on the Moon, while the Service Module would have been used for blasting off the Moon. However, as discussed earlier, an analysis of the problems of descending to the lunar surface led rapidly to the conclusion that an exceptionally large rocket would be necessary to put these three modules into a trans-lunar trajectory–the projected Nova vehicle.

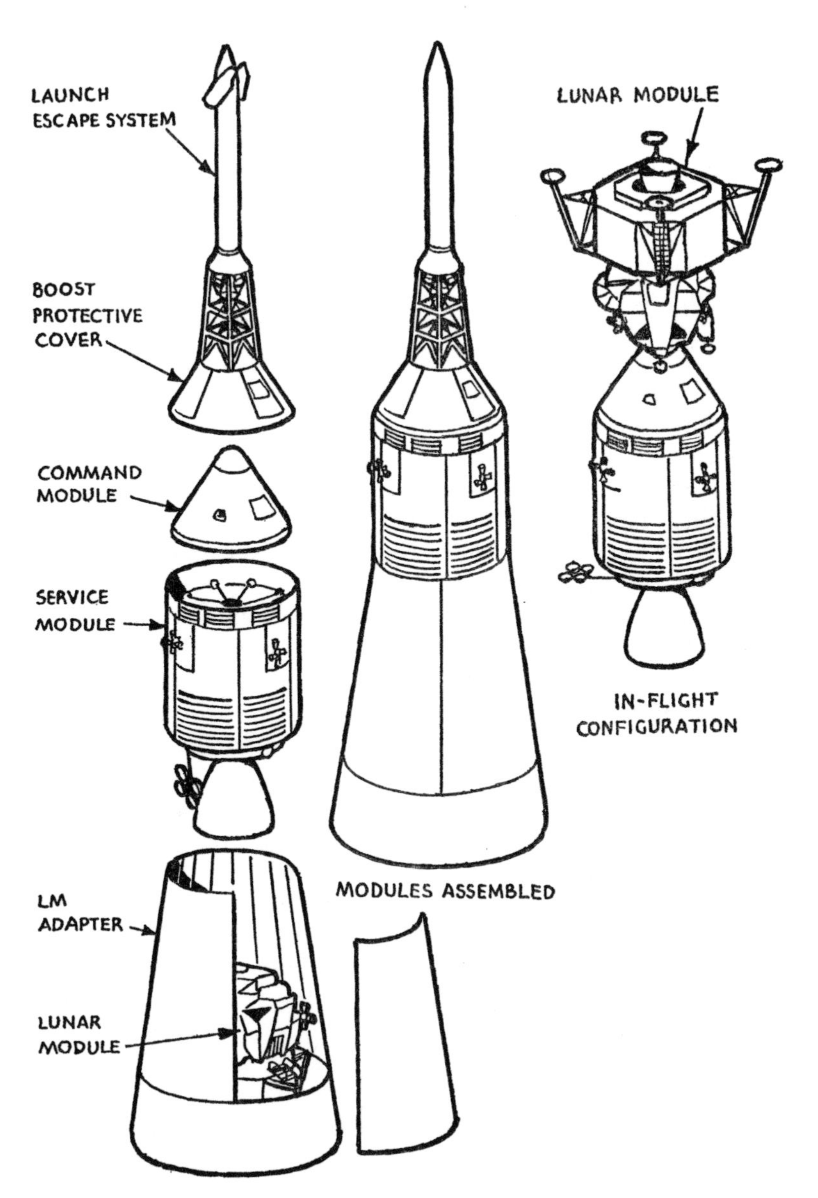

Fig. 8. The Apollo Spacecraft Modules shown separately, assembled for launch and as assembled in flight.

A lower 'total energy' procedure was therefore followed, in which an auxiliary vehicle would be launched with the Apollo spacecraft from Earth. This would accompany the main spacecraft all the way to the Moon, where it would be used for the actual descent from lunar orbit to the surface of the Moon, while the remainder stayed in orbit. This vehicle was originally known as the Lunar Excursion Module (LEM), but it is now termed just the Lunar Module or LM. Again, with an eye to minimising the total energy requirement for the mission, it was decided that only two of the three-man crew would descend to the Moon's surface in this Lunar Module. This module is itself divided into two parts, the lower or descent stage containing the propulsion system appropriate to retarding the orbital speed round the Moon, controlling the descent and ultimately effecting the soft landing. This section is then used as the launch pad for the upper or ascent stage which, by its own (yet another) propulsion system, lifts off from the lunar surface, accelerates into lunar orbit, and then makes a rendezvous with the parent Command Module, left in orbit. The two 'lunarnauts' then transfer back into the Command Module, jettison the Lunar Module and return to Earth.

This three-module concept will be discussed in more detail in the following chapters, but before passing on, it is worth mentioning the tower which can be seen in Fig. 8 on top of the Command Module. This is the Escape Tower, which is attached to a protective shield on the conical surface of the Command Module during the early part of the Earth lift-off. It is a larger and more sophisticated version of the tower used on the early Mercury spacecraft, and its purpose is to allow the astronauts to escape from the Saturn V launch vehicle in the event of an emergency by igniting a rocket motor, which accelerates the whole Command Module away from the other modules and the main rocket vehicle. This could be needed if serious trouble were to occur in the first or second stage burning periods of the Saturn V, which, of course, contains thousands of tons of liquid propellent, equivalent to a high explosive.

5

The Command Module

As the name implies, the Command Module is that part of the overall spacecraft which is the command centre for the mission. The module houses the three astronauts together with the life support equipment appropriate to the re-entry phase into the Earth's atmosphere–the only period when the Command Module is on its own, as distinct from the remainder of the mission when it is supported by, and attached to, the Service Module. Fig. 9 indicates the general truncated cone configuration of the Command Module with the ablative heat shield on the bottom which is designed to burn partially away during re-entry. A thinner coating of ablative material covers the other surfaces of the module.

Other main sub-systems of the Command Module, in addition to the life support system, comprise the guidance and navigation facility, a complex rocket propulsion system involving twelve small rocket engines which is used for stabilisation and attitude control in three planes, but which is not designed to provide for main velocity changes. There is also, of course, an electrical power system to feed the general services and, naturally, a comprehensive communications system. There is a separate sub-system for Earth landing and another for launch escape.

The Command Module is divided into three compartments–the aft compartment, the crew area and the forward compartment. The forward compartment is at the apex of the spacecraft and is a fairly small area, whereas the crew compartment occupies most of the centre section of the structure. The aft compartment is again fairly small and is located around the periphery of the module, near the base.

In the launch and re-entry phases, the Command Module is so oriented that the flattish heat shield end is pointing towards the Earth, and at these times the astronauts are on their backs on couches which are installed so that they face

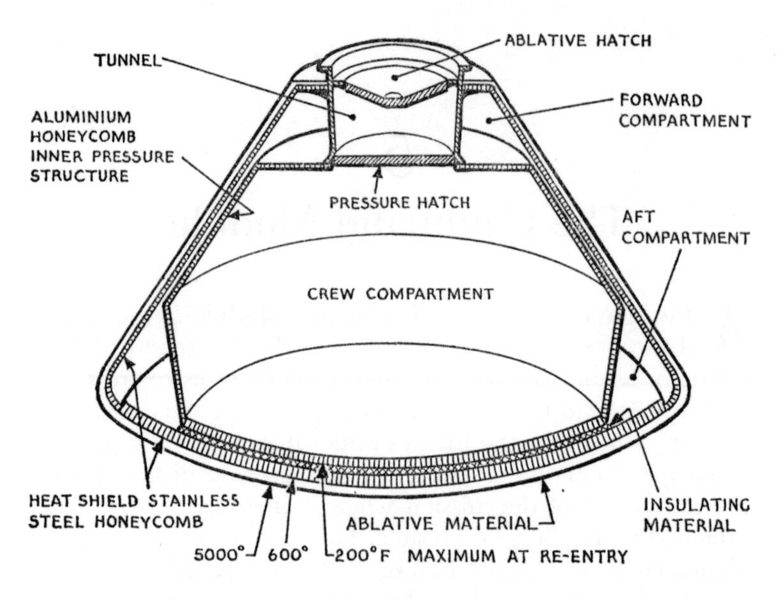

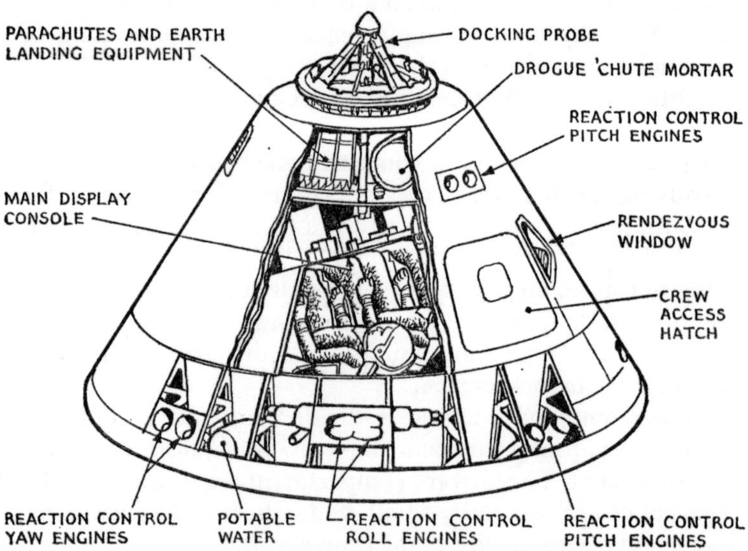

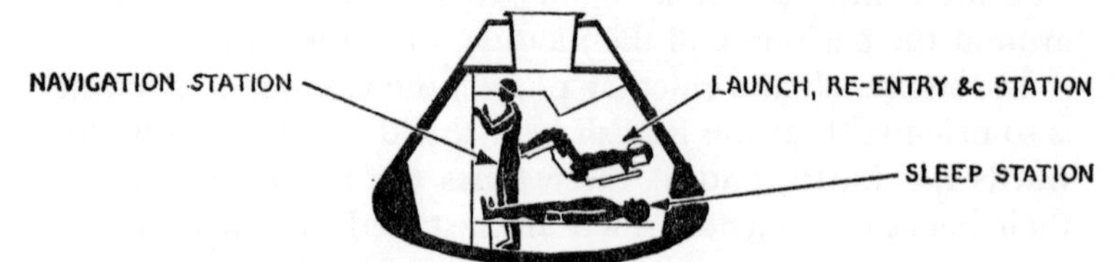

Fig. 9. *Top:* Section through Command module structure.
Centre: General View of Command module.
Below: Astronaut Positions.

the apex of the module. In flight the general weightless condition means that the orientation of the spacecraft makes little difference to the crew, except at those critical times when the craft's propulsion engine (housed in the Service Module) needs to be ignited for velocity changes, or when the guidance and control system is being used to establish navigation fixes on stars–and, of course, during the rendezvous and docking operations with the Lunar Module in Moon orbit. It is, however, convenient that in flight the module should normally be oriented so that the apex is forward along the line of flight.

The astronaut on the left-hand couch of the three is the spacecraft commander and he normally operates the spacecraft flight controls in addition to his general command duties. The centre couch is the place for the Command Module pilot, whose principal task is guidance and navigation although he can control the spacecraft when necessary and, indeed, in the actual lunar landing phase he is the member of the crew who remains in the Command Module while the other two descend to the lunar surface. The astronaut on the right-hand couch is the Lunar Module pilot, but in addition he is concerned with the management of the spacecraft sub-systems. The centre couch can be folded and, with the seat portion up, two astronauts can stand at the same time. Indeed, all the crew members can leave their couches and move around in conditions which are not unduly cramped, bearing in mind that the diameter of the module is 12 ft 10 in externally at the base with a height of 10 ft 7 in. Each man has about 73 ft^3 of space–about that available in a telephone kiosk. The weight at launch, including the crew is some 13,000 lb which is reduced to about 11,700 lb at splashdown.

There are some missions or phases of missions when it is convenient for all three astronauts to sleep simultaneously, but generally two sleep at a time, when they move to two sleeping-bags which are mounted under the left- and right-hand couches. These bags are attached to the main structure and have restraints so that the crew members can sleep in or out of their space suits.

The cabin is normally pressurised to about 5 lbf/in^2–about a third of sea-level atmospheric pressure–and is maintained at around 75° F (23° C). At launch the astronauts are in their pressure suits, but once they are in Earth orbit, or are successfully injected into a trans-lunar trajectory–depending upon the mission–they are able to get out of these suits and spend the rest of the 'cruise' time in what is referred to as the 'shirt-sleeves' condition. They don their suits again for the docking and transfer manoeuvres, but for most of the time they have reasonable freedom of movement.

Apart from the normal technical aspects and the sub-systems mentioned, the bodily needs of the crew need to be attended to, and these are dealt with by the food, water, clothing and waste management equipment which is packed into bays lining the walls of the spacecraft. These matters will be discussed in a later chapter, but it is worth noting here the great ingenuity which has been exercised in fitting a mass of mixed equipment into a small, and what must be considered a rather unusual, shape.

Although the spacecraft is complex and has about two million parts, it has been designed so that one astronaut alone can return it safely to Earth in the event of difficulties arising with the rest of the crew or in the event of the lunar landers not returning.

In its general configuration there are two hatches–the main crew hatch in the side of the module through which entry is made on the launch pad, and a further hatch at the apex through which the astronauts climb into the Lunar Module. There are five main windows–two rendezvous windows and two side windows, and an integrated pair of windows for the sextant and scanning telescope for the navigation system.

Some difficulty had been experienced during the trials programme of Mercury and Gemini, and even with Apollo, through occasional fogging of these windows. In the Apollo 8 mission this turned out to be the result of 'outgassing' from the RTV sealing compound round the edges of the triple-layered windows and a modification programme was put in hand for subsequent spacecraft. Finding the source of this

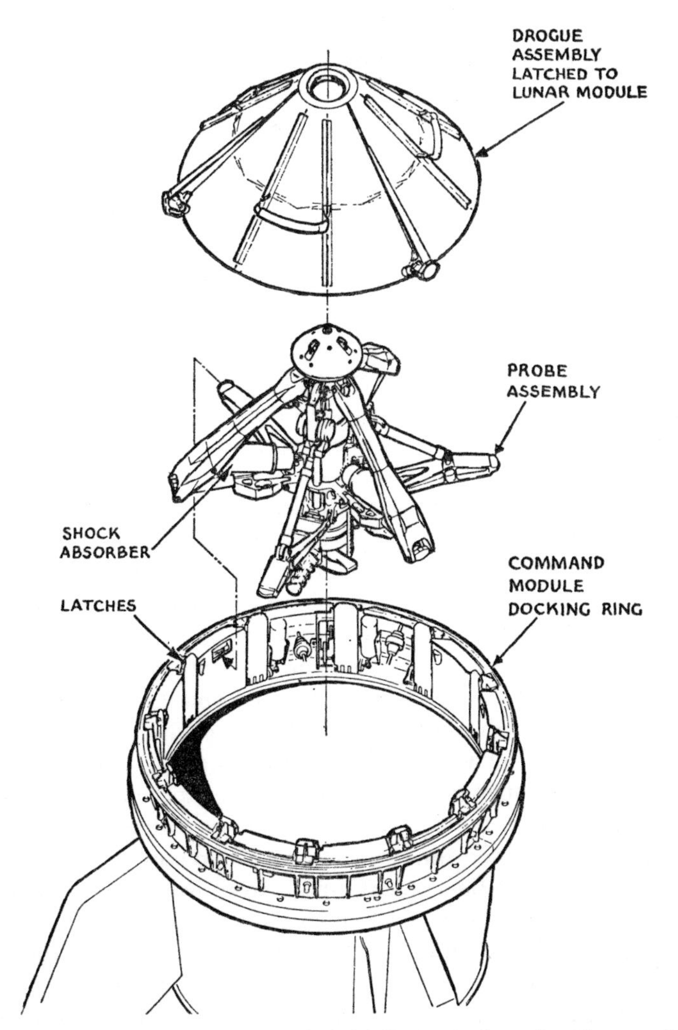

Fig. 10. The Probe Assembly is initially latched to the Command Module Docking Ring, and, in docking, is guided into the Drogue Assembly latched to the Lunar Module. The two modules are then drawn together until locked. The probe and drogue are then unshipped to clear the tunnel for astronaut passage.

trouble at last is very important for, bearing in mind that clear vision is required especially through the sextant window, this fogging if it had persisted could be more than an irritation–it could have been a fundamental hazard to the success of the flight.

Around the periphery of the spacecraft are located the twelve reaction engines used for yaw, pitch and roll control, the elongated apertures indicating the efflux ducts for these engines. They form the Command Module attitude-control system used to orient this module.

The construction of the Command Module is interesting in that it has an inner pressure-shell structure and an outer heat-shield structure. One feature of these two separate structures is a series of four crushable ribs on one side of the rear section between the inner and outer shells. The descent attitude of this module on impact is at an inclination of some 27 degrees to ensure that the impact on the sea is taken on this lower section. The ribs are of crushable material and comprise the external shock absorbers, the whole of the inner cabin shell moving down somewhat when the ribs are crushed. Further attenuation of the impact on the astronauts themselves is given by hydraulic dampers attached to each of the couches, between the couch and the inner structure.

This inner structure is of aluminium sandwich construction–adhesively bonded honeycomb core with an outer face sheet and a welded inner skin, the honeycomb varying from ¼-in thickness on the conical surface to 1½ in over the circular base. This inner shell is, of course, pressurised and contains the normal breathable atmosphere. The outer structure is made of stainless steel brazed honeycomb sandwiched between two alloy steel face sheets, and this again varies in thickness between ½ in on the sides and 2½ in on the base. Insulating material to provide additional heat protection for the crew is located between the inner and outer shells.

The structure, in addition to providing normal rigidity and strength for the internal pressure, must also provide thermal protection. During the launch phase aerodynamically generated heat is absorbed principally by a boost protective cover, which is of fibre-glass reinforced material covered with cork. This fits snugly over the Command Module during the lift-off and first part of the flight. This cover is attached to the launch escape tower and, during a normal mission, is discarded at nearly 300,000 ft altitude.

The heat shield on the module base is one of the most interesting parts of the spacecraft. The outer layer is ablative, about 2 in thick, and is made of a phenolic epoxy resin, a type of reinforced plastic, which is intended to burn away to some extent during re-entry. The material turns white hot, chars and then melts, flying off the spacecraft and taking with it the aerodynamic braking heat which does not then penetrate to the interior of the spacecraft. This ablative material is continued up the sides of the Command Module but decreasing in thickness until it is only ½ in thick at the apex.

During the re-entry phase this heat shield ablative material reaches about 5000° F (2750° C), and this has a serious effect on the radio frequency characteristics of the communications system, as the plasma sheath that is formed by the super-heated air around the spacecraft causes a radio blackout, which lasts for a few minutes while the module is coming through the outer fringes of the Earth's atmosphere. Although some attempts have been made to try and reduce this effect–for example, by injecting water into the plasma sheath during the appropriate part of the re-entry phase–little improvement has been achieved and the crew now accept the occurrence of a radio blackout at this time. It is not, in fact, unduly serious as the spacecraft is essentially dependent for survival on its internal functioning and instrumentation at this period.

At this point it is worth pointing out that while there is a tremendous dependence by the crew on the computer and communication and control facilities established on Earth, there are certain functions, particularly in the guidance system, that are fundamentally dependent upon crew action, so that ultimately the safety of the crew and their return to Earth depends a great deal on their own efforts. This philosophy has evolved over the years from just before the Mercury programme when it was thought perhaps essential to ensure fully automatic procedures, which could take over the control of the spacecraft in the event of failure of the crew. However, in the early days of Mercury it became apparent that the crew could and should play an integral part in the conduct of their own mission. In the words of Colonel Borman, the commander of the Apollo 8 spacecraft which made the first manned flight round the Moon in December 1968, 'the approach now of the astronauts to controlling the spacecraft is more allied to considering it as an aircraft than as a super type of guided missile'. Clearly, there are many functions that cannot be carried out by the crewmen, but by the time the reader has finished this book it will be apparent that the safety of the crew and the success of the missions depend basically on the skill and training of the astronauts.

The picture emerges, then, of a complex Command Module as the focal point of the whole Saturn V/Apollo vehicle, with the three astronauts an integral part of the system. Such has been the crew performance in the first few years of the space programme–including that of the American's Russian counterparts, the cosmonauts, with Valentina Terishkova, the only woman so far to orbit the Earth–that there is a tendency to consider them as not just supermen, but as tireless, nerveless supermen. Several events have, however, occurred which remind us that they are still men.

To begin with we can note that great attention has been paid to maintaining a proper environment in the spacecraft and in providing in Apollo (although not in Gemini and certainly not in Mercury) a reasonable degree of freedom to move about. The crew members carry out limited exercises, for example, by pulling on a rope with their arms against a frictional resistance, regularly during the course of each day in space. In general, the arms are well used and it is the leg muscles which tend to deteriorate. Indeed, after his first flight in Gemini and 14 days in orbit, Colonel Borman was by no means sure that he and his fellow-astronaut would be able to walk at the end of the mission. In fact, they were both able to–but only just; and this demonstrated the need for properly planned exercise during the course of a flight.

We know, of course, that the human body can conjure up

reserves of energy when necessary. However, the problem of tiredness and fatigue on lunar missions is a very real one, and the awareness of this problem in its mental and physiological senses derived from the Apollo 8 flight must have been most useful in planning to minimise fatigue in later missions, by introducing 'forced' relaxation periods in readiness for later tasks which would demand great effort.

Closely associated with the comfort and health of the crew is the waste management system. Solid bodily waste is, in fact, stored in sealed plastic containers after the addition of bacteriological powder, while urine is dumped overboard through a small diameter nozzle, heated to prevent the formation of ice. Excess water from the fuel cells is similarly dumped overboard–or evaporated to provide a cooling function–and these two liquids are the only materials to leave the spacecraft during the course of a voyage, other than the efflux of the rocket engine and attitude thrusters. The liquids, of course, immediately freeze and, with a small exit velocity from the spacecraft, slowly move away from it. These ice particles are visible and Colonel Borman remarked that on the 14-day-long Gemini flight in which there was some aspect of boredom, they saved the urine dump until sunset, when the lighting effect produced a spectacular sight as the coloured ice particles moved away from the spacecraft!

It is not generally realised that with the outgassing from materials in the spacecraft construction, the efflux from the thruster nozzles and the water dumping, quite apart from the molecules of the atmosphere trapped on the spacecraft, a significant amount of matter can be carried along in a cloud around the spacecraft. Consequently, near the craft there is by no means the hard vacuum of space, and one would need to move some distance away from the spacecraft to avoid incidental collisions with the molecules of this matter. Scientific experiments have, therefore, to take this into account.

Before we leave the Command Module structure let us look briefly at the two hatches. As will be discussed later, the side hatch had to be completely redesigned as a result of the disastrous fire during a Command Module test in 1967 when three astronauts lost their lives. The original hatch, while simple in concept, took over a minute to open through involving a number of sequential functions. The present side hatch is a single integrated assembly which opens outwards and has primary and secondary thermal seals. It is 29 in high and 34 in wide, and weighs about 225 lb. It can be opened in a few seconds by a handle which the crew man pumps back and forth to open eventually the twelve latches round the periphery. The hatch is designed so that the internal pressure exerted against the hatch serves only to increase the locking pressure of the latches.

The forward hatch is used in docking with the Lunar Module. It is a combined pressure and ablative hatch and is mounted at the top of the 'tunnel' through which the astronauts crawl when the two craft are docked. The hatch is 32 in diameter and weighs about 80 lb, while the exterior upper side is covered with ½ in of insulation and a layer of aluminium foil. It is a more simple and perhaps less critical hatch than the one in the side, in that its operation can be coped with slowly as part of the general docking and undocking procedure; it has no emergency exit function as does the more critical side hatch.

6

The Command Module Systems

THE success of the Command Module depends, of course, on the many separate systems and sub-systems, which will now be given some attention.

First, some comments about the displays and controls. There are, in fact, twenty-four instruments displayed in the Command Module panel, as well as forty mechanical event indicators and seventy-one lights. There are also 566 switches. The majority of the controls and displays are mounted on the main display console which faces the three crew couches and extends on both sides of them. (See Plate 5A.) The console is nearly 7 ft long and 3 ft high, with two wings on each end. Most of the guidance and navigation equipment is in the lower equipment bay at the foot of the centre couch, and this includes the sextant and telescope, which are operated by an astronaut standing and using the simple restraint system. The rotation and translation controllers, which are used for changing craft attitude, for aligning the propulsion engine (of the Service Module) thrust vector, and for translation manoeuvres, are located on the arms of two of the crew couches. A rotation controller can also be mounted, when required, at the navigation position in the lower equipment bay. Clearly the Command Module pilot in the centre couch faces the centre of the console and can reach all the necessary flight controls.

The controls can all be operated by astronauts wearing pressure-suit gloves, and there are four basic types—toggle switches, rotary switches with click stops, thumb-wheels and push buttons. Locks and guards are provided for critical switches which could cause problems if operated inadvertently although, of course, nothing is infallible. Indeed, although it proved not to be serious in the event, one slightly disturbing thing of this nature occurred during the Apollo 8 flight. An incorrect instruction was given by one of the crew members via the onboard computer, which caged the gyros of the inertial navigation system when the spacecraft was a day out from Earth. Fortunately, although their initial space reference had thus been lost, the navigation and guidance system was of such excellent design that they were able to uncage the gyros and reset them. The point to be noted, however, is that all systems are open to human error finally, however well trained the operators.

Electrical power is, naturally, one of the major sub-systems and this system provides electrical energy via power conversion and distribution units throughout the mission. Direct current power is supplied by ground support equipment prior to launch, the changeover to onboard power being made at lift-off. The electrical power sub-system furnishes drinking water for the astronauts as a by-product of the fuel cell-type power plant. In addition to the oxygen and hydrogen tanks associated with the fuel cells, there are three silver oxide zinc batteries and inverters, which are located in the lower equipment bay.

The environmental control sub-system provides a controlled environment for three crew members for up to a maximum of 14 days. This system has been designed so that the crew need spend the minimum time operating it, and its objective is to remove carbon di-oxide and other unwanted odours from the Command Module cabin, to provide fresh oxygen and water, and to make provision for the disposal of waste material and the dissipation of excessive heat both from the cabin and from the electronic equipment.

This sub-system is concerned with three major elements–oxygen, water and a water-glycol coolant–though these are inter-related and integrated with other sub-systems. These elements provide the major functions of spacecraft atmosphere, thermal control and water management through three sub-systems, while a fourth is the pressure-suit circuit and a fifth sub-system is that of post-landing ventilation, which provides outside air for breathing and cooling after the Command Module has splashed down. The whole environmental control unit is quite compact, measuring 29 in long, 16 in deep and 33 in at the widest point. It is mounted in the left-hand equipment bay.

At lift-off the cabin atmosphere pressure is a little above atmospheric, but it falls as the spacecraft rises and when it reaches 5 lbf/in^2 automatic equipment begins operating to maintain nominally this pressure throughout the flight. As the cabin pressure decreases, oxygen from the pressure-suit circuit is dumped into the cabin so that over the launch phase there is a constant differential pressure between the suit and the cabin space.

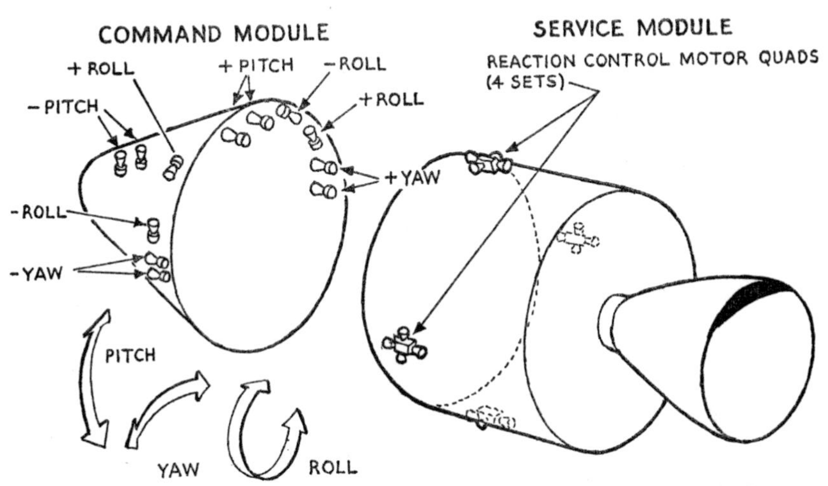

Fig. 11. Reaction Control Thrust Engines for the Apollo Command and Service Modules. The systems are duplicated, one of any pair being sufficient for its task.

The system can deal with high leak rates and can cope even with a ¼-in diameter hole in the cabin wall, which might be caused by a micro-meteor impact, for example. In this situation, 3·5 lbf/in^2 of pure oxygen atmosphere can be maintained in the cabin for more than 15 minutes, which can sustain the astronauts long enough for them all to don their space suits and tap into the pressure suit circuit of the environmental control system. The oxygen supply to the cabin can then be cut off to stem wastage of what is, after all, a very precious commodity in space.

Another important Command Module system is that for the control of the reaction thrusters. This is used principally after the separation of the Command and Service Modules, which occurs about 15 minutes prior to the Command Module's re-entry into the Earth's atmosphere. It is also used for certain Abort Modes. The object of this system is to provide attitude and roll control to maintain the Command Module in the proper attitude prior to and at re-entry, both before encountering aerodynamic forces and after to provide the roll position control which is used to fly the spacecraft to the final point of impact, as is explained later.

The twelve engines in the Command Module Reaction Control system consist of two independent systems each containing six engines. Each system has associated with it helium pressurisation and propellent storage tanks, and dump and purge systems. The two systems can operate in tandem, though either system alone is adequate to provide the forces necessary for the various manoeuvres, and normally only one is used, the other being 'back-up'. All these small rocket-thruster engines are located outside the crew compartment, ten in the aft compartment and two in the forward compartment, and each produces about 93 lbf thrust by reaction. The propellent is the same in both systems–monomethyl hydrazine fuel and nitrogen tetroxide oxidiser – and is hypergolic, so that no igniters are required. The helium is used to pressurise the propellent in the tanks and, as this takes place in the weightless condition of 'free-fall', the helium is fed to positive expulsion bladders, which blow up like balloons and force the liquids out of the tanks. These hypergolic propellents could constitute a hazard at splashdown, so that during the final parachute descent, when the use of the reaction control system is over, these propellents are burnt off in the atmosphere.

Re-entry into the Earth's atmosphere takes place at a nominal angle of 6 degrees above the horizontal with an absolute tolerance of plus or minus 1 degree up or down from this angle. The re-entry angle is determined such that at the nominal 400,000-ft altitude following which the first phase of re-entry takes place, there is a gradual heating up of the heat shield and deceleration of the vehicle, which soon enters what has been called the S-band blackout–which means that radio waves of between 1550 and 5200 megacycles/second cannot get to or from the spacecraft. A few seconds after this the accelerometers indicate a deceleration of 0·05 *g*, when an initial roll rate is introduced through the Command Module reaction engines in order to reduce dispersion and this is maintained until an altitude of around 200,000 ft is reached, by which time the deceleration has built up to 4 *g*.

At this point the Command Module is rolled, so that the lift force moves to an upwards direction and begins to flatten out the re-entry trajectory and, indeed, the spacecraft begins to climb away from the Earth slightly. This lift force is achieved by arranging for the centre of gravity of the Command Module to be offset from the centre line; the spacecraft consequently tends to come in tilted some 20 degrees to the line of flight, so that as well as the braking force along the flight line there is also an aerodynamic lift force at right angles and in the line of the tilt. This lift force is used during this phase to keep the spacecraft out of the denser atmosphere and to maintain no more than 4 *g* deceleration and, after sufficient speed and thus energy has been lost, the vehicle is rolled again so that this 'lift' is in a downwards direction. The flight path curves back towards the Earth and soon after this the C-band (3·9–6·2 gigacycles/second frequency) and then the S-band blackouts end.

The vehicle is moving back slightly towards the Earth again at this phase and it is rolled finally in a relatively slow forward speed with an increasing descent angle until the Command Module guidance terminates at about 60,000 ft. Soon afterwards at 24,000 ft the drogue parachutes are deployed followed by the three main parachutes at 10,000 ft. These parachutes are stowed at the forward end of the Command Module round the forward hatch tunnel and the main chutes are rigged so that the craft touches down with a 27½ degree tilt for controlled impact.

Returning to the control aspects of re-entry, there are a variety of indicators in the control panel of the Command Module. In addition to normal altimeters, there are two flight director attitude indicators, which are spheres mounted such that they can indicate the actual and the required attitude in pitch, roll and yaw, and there is an entry monitor system display panel, which is used during the critical early phase of re-entry. This display panel has a grid marked on it such that during the initial pull-up to maintain a maximum of 4-*g* deceleration, should the automatic system be in error and the pull-up be maintained for too long, the flight computer shows that the path is moving too high above a set of parallel lines which slant upwards on the right side of the display. These lines represent the limits of the climb path which, if exceeded would mean that the spacecraft would bounce off the outer atmosphere and skip off into a farther, possibly disastrous, space orbit. Consequently, if a point is reached when the lines become parallel, immediately the Commander should punch a knob which switches from automatic to manual control, and manually roll the spacecraft so that the lift vector is in a downwards direction to decrease the flight path. It is of interest that on the Apollo 8 circumlunar mission, Colonel Borman was indeed ready to take over manual control at re-entry, because the climb path did achieve a parallel with the limit. However, the system was, in fact, working perfectly and the flight continued successfully on automatic control.

As always in space activities, the communications equipment is very important and this is certainly the case with the Apollo spacecraft. A communications sub-system provides voice, television, telemetry and tracking and ranging communication between the spacecraft and Earth, between the Command Module and the Lunar Module and between the spacecraft and the astronauts, who might be wearing the portable life-support system on the Moon's surface–equipment which is discussed more fully in a later chapter.

Obviously, communication is also provided between the astronauts in the spacecraft, who might be in their pressure suits, and this is achieved by normal light-weight headsets used for all voice communications. There is a comprehensive data-gathering system monitoring the spacecraft structure, the many sub-systems, together with biomedical television and timing data, which is transmitted back to Earth. The radio frequency equipment used consists of two VHF/AM transceivers (transmitter-receivers) in one unit, the unified S-band equipment (primary and secondary transponders for radar tracking and an FM transmitter), primary and secondary S-band power amplifiers, a VHF beacon for recovery purposes, an X-band transponder (for rendezvous radar) and the premodulation processor. These units will not be considered in detail here, but further information about the principles of the tracking, navigation and communications systems is given in a later chapter.

The term S-band refers to radio waves in the frequency range 1550–5200 megacycles/second. Other letters signify other ranges. The term 'unified S-band' has been adopted by NASA, because the old practice of using quite different frequencies for voice, television and data transmission has been replaced by a system in which all are considered as data, and use the same transmitters and receivers. All has been 'unified' into a single radio telecommunications link.

Finally, we can note that during the course of the flight to the Moon, except when precise movements are being used for navigation or observation purposes, the spacecraft undergoes what is referred to as a 'barbeque' motion, whereby it is rolled at the rate of one-tenth of a degree a second about its longitudinal axis, the roll being controlled by the Service Module thrusters. The object of this is to expose gradually the different sides of the spacecraft to the Sun on one side and to the cold of space on the other side, thus easing the environmental temperature-control system's work, which would otherwise have to cope with a heat sink and a heat source at constantly diametrically opposite sides of the spacecraft.

What has been discussed so far only briefly touches some of the more interesting aspects of the Command Module and its systems; but from this it will be seen that it is probably the most compact and complex piece of equipment that has ever been produced by man. Certainly, with a mission that requires the maximum reliability, the demands made would have been beyond the wildest possible imagination of engineers even a few years ago. For it is one thing to design a spacecraft that can go to the Moon, but it is another to ensure that the reliability and integrity of its complex systems is at that high level necessary before men can be committed without unreasonable risk.

7
The Service and Lunar Modules

THE Service Module is an integral part of the Apollo spacecraft in that the Command Module cannot operate without it, except for the small time of re-entry. The Service Module provides the main propulsion services and supplies most of the Command Module's consumables, such as propellents, hydrogen, water and oxygen, throughout the lunar

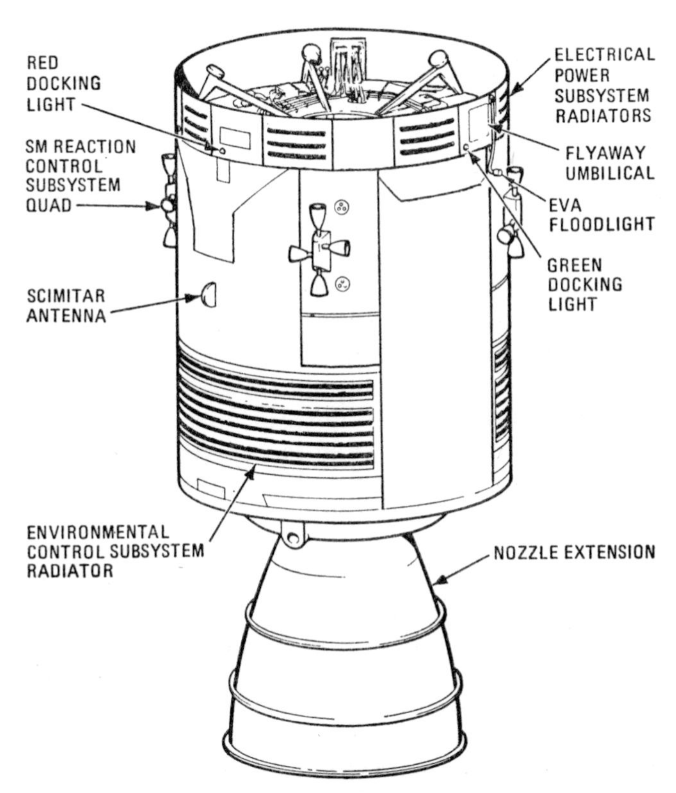

Fig. 12. The Apollo Service Module.

flight. It is, of course, not manned, and it is finally separated from the Command Module just before re-entry, when it burns up during its own uncontrolled re-entry.

This module is a complete rocket stage in itself, as seen in Fig. 12 and it is some 24 ft in length and 12 ft 10 in in diameter to match up with the base of the Command Module. When fully loaded with propellents and consumables it is 55,000 lb in weight, though when dry it is only 11,500 lb. These are, however, only 'round figures'; they vary for different missions and, over a period of time, the weights of all the spacecraft modules have tended to rise, a situation which has been matched by gradual up-ratings of the launch vehicle rocket engines.

The propellents carried in the Service Module feed the main Service Propulsion Engine, which has a thrust of 25,000 lbf in vacuum. The engine itself is 3 ft 5 in long, with a radiation extension nozzle of 9 ft 4 in. This main engine is gimballed for control purposes, and its propellent is 50% hydrazine and 50% unsymmetrical dimethyl-hydrazine, with an oxidiser of nitrogen tetroxide–these, again, being hypergolic so that they ignite on contact.

This propulsion unit is one of the key parts of the whole spacecraft and the fundamental nature of its reliability in making a mission successful was brought out publicly for the first time by the Apollo 8 flight, when it was relit a number of times–to decelerate the spacecraft into lunar orbit, to correct this to a near circular orbit, and to impart the final escape velocity to bring the craft back to Earth, as well as for course corrections.

Because of the characteristics required of this engine, a massive development programme was undertaken by its makers, the Aerojet General Corporation of America, and not only were the critical parts of the pumps, valves, tanks, etc., duplicated, but the especially vulnerable parts were quadruplicated. The only parts which were not duplicated were the injector and the main thrust chamber of the engine itself. As a result these items were probably some of the most carefully inspected and tested pieces of equipment ever produced by industry. In addition, also, to the normal manufacturing and development checks, the stability of the propellents burning in the chamber were checked thoroughly by exploding as many as 187 small charges inside the chamber in various tests, attempting to see whether burning instability could be created by such anomalies, because instability in burning could lead to faulty performance of the engine in flight. These tests produced no adverse effects and its performance has been vindicated by the flights to date.

Around the outside of the cylindrical section of the Service Module are the four 'quads'; these are units each of four reaction control engines each with a thrust of 93 lbf. They provide the thrust for three-axis control and stabilisation of the whole spacecraft until the time when the Service and Command Modules separate. This reaction control system is used also to provide an ullage space in the main propellent tanks. This means that after a prolonged period of weightlessness, the thrusters provide a small accelerating force which settles the fluids at the bottom of the tanks before the main propulsion engine is started up. While this engine is actually thrusting, the small reaction engines only control roll of the craft, pitch and yaw control being exercised by gimballing the Service Propulsion Engine. The reaction control system can also provide velocity changes for spacecraft separation from the third stage of the launch vehicle during High Altitude or Trans-lunar Injection Aborts, and it can be used for a number of other manoeuvring operations where only small changes in spacecraft velocity are required.

The reaction control engines can be pulse-fired in bursts to produce short thrust impulses or they can be fired continuously over periods to produce steady thrusts. These engines are able to be fired the many hundreds of times that may be required over the course of a mission. The main Service Propulsion Engine may only have to be relit a few times, but it has provision for up to fifty starts.

The Service Module structure itself is relatively simple and consists of a centre section surrounded by six sectors shaped like pie slices. The construction is of various materials–honeycomb panels for strength and lightness, and solid aluminium alloy radial beams, for example–and there is an aft heat shield which surrounds the Service Propulsion Engine to protect the remainder of the module from the engine's heat during thrusting. The six radial sectors contain various items, such as the oxygen and hydrogen tanks, the fuel cells, helium pressurisation tanks, and so on, as well as the propellent and oxidiser tanks for the main engine and for the reaction control engines.

On the outside of the Service Module are located the space radiators for both the environmental control and electrical power sub-systems. There are also various aerial antennae, umbilical connections and lights. Particularly apparent is the S-band high-gain antennae with the four 31-in diameter reflectors surrounding an 11-in square reflector. At launch this is folded down parallel to the Service Propulsion Engine nozzle, so it fits within the spacecraft envelope near the Lunar Module Adapter. After the separation of the Command and Service Modules from this adapter, the S-band antennae unit is deployed at right angles to the Service Module.

The life of the Service Module finishes just before re-entry and at this time a number of complex events occur in rapid sequence as the Command and Service Modules separate. These not only include the physical separation of all the connections between the modules–and they are many–but also the transfer of electrical control, and the firing of the Service Module's reaction control engines to increase the distance between the modules and so on. It is also necessary to ensure that the separation causes no subsequent effects within the Service Module which could create a hazard to the final phase of flight of the Command Module.

While it is not as complicated as some other parts, it can be seen that the Service Module is a key element in the whole Apollo story.

The third of the modules carried on the journey to the Moon is the Lunar Module and this is housed behind the Service Module in a section called the LM Adapter during launch and injection into trans-lunar trajectory. Of all the modules, this is the most unorthodox in appearance and it is, perhaps, not surprising that it has been referred to colloquially as the 'Moon bug' or 'Spider'. Its bizarre appearance arises from the fact that as it never has to operate in an atmosphere, it does not have to take on a streamlined shape. Not only do the various parts only have to be given a light protective covering, but they do not have to be marshalled to fill a long cylindrical envelope. The result is a rather 'shapeless', squat spacecraft. So different, in fact, were the requirements

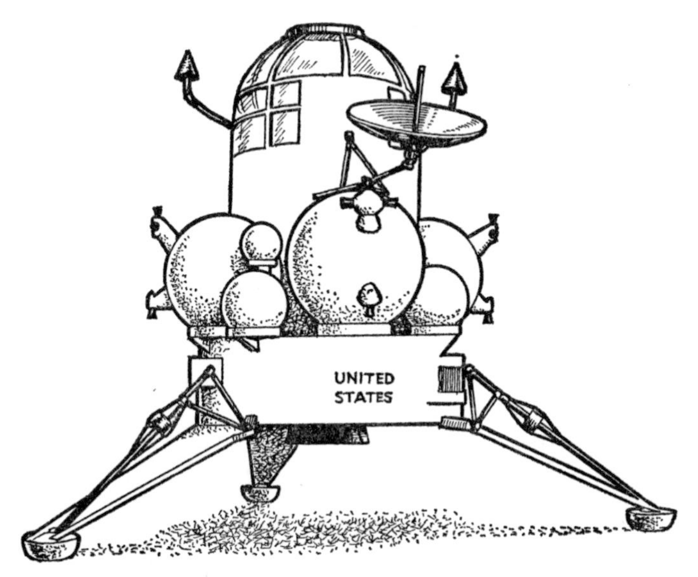

Fig. 13. A very early idea for a Lunar Module.

from other manned spacecraft, that the Lunar Module had to evolve slowly over a number of years of development.

Figure 13 shows an early concept by NASA of the kind of configuration they had in mind. This had a base, with three legs similar to those of the Surveyor unmanned craft, a central astronaut cabin and a number of exposed spherical tanks surrounding it for the consumable liquids and gases. As the result of successful tendering, in November 1962, the Grumman Aircraft Engineering Corporation were given the job of designing and developing the Lunar Module. The contract included also a number of 'boiler plate' models for rocket-launch tests, and a number of flight simulators, some for experimental work and others for training purposes. These will be considered later.

Through co-operation between Grumman and NASA engineers the configuration eventually changed into that shown in Fig. 14. It is a two-stage vehicle. The upper ascent stage comprises the unit which houses the two astronauts who are making the descent to the Moon, together with the propulsion system for taking off from the lunar surface and

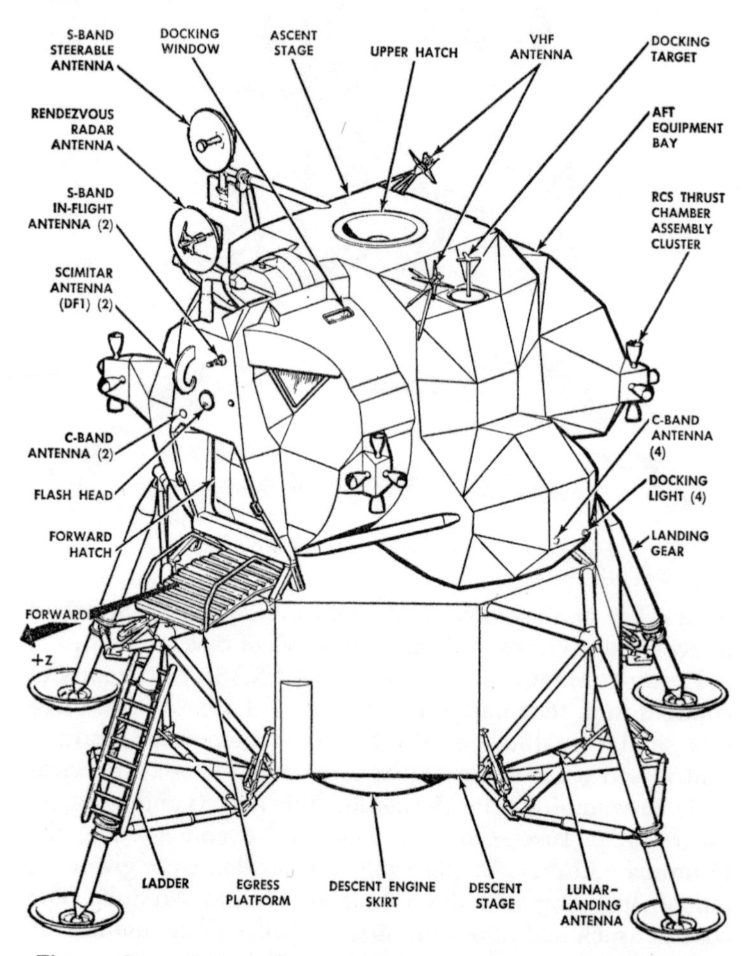

Fig. 14. General view of Lunar Module and names of exterior parts.

making a rendezvous with the Command Module left in lunar orbit. The lower descent stage of the Lunar Module is equipped with four extendable legs and its own propulsion system, this being used for propulsion away from the Command Module and for the descent to the lunar surface. The overall height of the Lunar Module is 22 ft 11 in with the legs extended, and diagonally across the landing gear it is 31 ft, though with the legs folded up its diameter is reduced sufficiently for it to fit into the LM Adapter on top of the third stage of the Saturn V launch vehicle. At the beginning of its mission–that is, when it leaves the Command Module–it weighs 32,500 lb together with propellents and crew, though its dry weight is only 9000 lb. The ascent stage is 12 ft 4 in high and 14 ft 1 in diameter, with a dry weight of 4850 lb.

The Lunar Module serves as living quarters and a base for operations on the Moon and, of course, it allows the astronauts to return to the Command and Service Modules waiting for them in orbit. It is designed to operate for 48 hours entirely on its own, though at the moment it is planned to have a maximum duration actually on the Moon's surface of 35 hours.

Of all the parts of the Apollo spacecraft this, perhaps, has been the one which has been most sensitive to weight increases and the makers, in the 6–9 months prior to its first full-scale test flight, had to implement extremely meticulous engineering programmes in order to reduce weight in every possible way. Fractions of an ounce were saved in as many places as possible in order to try to keep the weight below the specified limit, beyond which it would be impossible to operate this module successfully. This arises not just because of its actual carriage to the Moon–when as seen in Chapter 2 every pound soft-landed requires about 120 lb of take-off mass from Earth–but also because of the relation between the weight of the Lunar Module and the thrust available from the rocket engines during its descent and landing on the Moon.

The descent engine provides the power for the manoeuvres required to take the module down, and it is a controlled thrust gimballed engine, which has a range of thrust from 9710 lb maximum to 1050 lb minimum. This range is quite adequate because with the lunar gravity being one-sixth that of Earth the effective weight of the Lunar Module fully fuelled is only about 5500 lb. This engine is ignited for a short burn on leaving the Command and Service Modules, when the module coasts down, having decreased its velocity, to the perigee of its descent orbit which is about 50,000 ft above the lunar surface and up-range of the proposed landing site. The descent engine is again fired at that point to reduce the velocity still further for the landing. The following descent is automatically controlled to an altitude of a few hundred feet by the guidance navigation control system. During the final phase the two-man crew select a favourable landing site and, by manual control of the reaction control system jets (which are clustered at the four corners of the ascent stage) and the variable thrust descent engine, manoeuvre the craft rather like a vertical take-off aircraft into the correct attitude over the landing site. The thrust is then progressively reduced and the craft touches down. Sufficient fuel is carried for the astronauts to be able to inspect the site and move up to a thousand feet sideways so that they can land in the best position in the area.

The first thing they have to do after landing is to check that all systems are functioning correctly and that the spacecraft has come to rest in an acceptable attitude ready for take-off. The landing gear is of the cantilever type and the four legs are connected to outriggers that extend to the ends of the descent stage structural beams. Each landing-gear leg has a primary strut, a footpad and a drive-out mechanism with two secondary struts. All of the struts have at their ends crushable shock-absorbing honeycomb inserts and the primary struts therefore absorb compression loads with the secondary struts absorbing the tension and compression loads. The forward landing gear has a boarding ladder on the primary strut which is used to climb to and from the ascent stage forward hatch. As an example of the weight-saving which has had to be introduced, this boarding ladder has been designed to be strong enough to be used in the reduced lunar gravity, but not strong enough to be used by a man in Earth gravity.

The take-off from the lunar surface, using the descent stage as a platform or launch pad, is satisfactory providing the craft has come to rest within $12\frac{1}{2}$ degrees of the horizontal. This is one reason why the astronauts have to be careful in

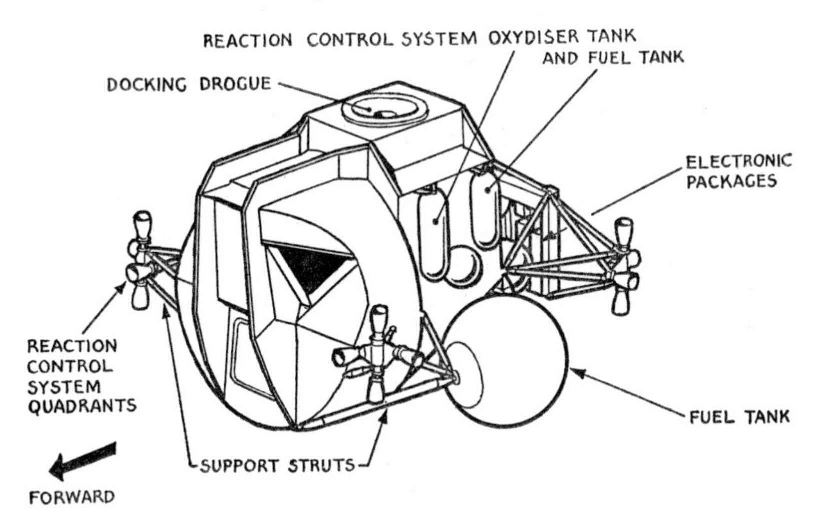

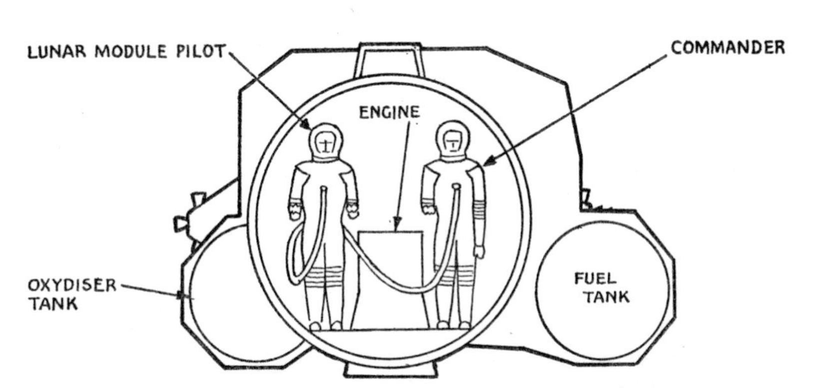

Fig. 15. *Top:* Basic Lunar Module Ascent Stage parts before outer covering is added.
Below: Front view of Lunar Module Ascent Stage showing crew positions and unsymmetrical tank arrangement.

their final choice of landing site–it must be a fairly flat and even surface, otherwise they might not be able to take off again.

The ascent stage propulsion engine uses a fixed constant-thrust engine with associated helium pressurisation and propellent components. Since this engine is not steerable the equipment in the ascent stage is positioned so that the centre of gravity is as near as possible to the line of thrust of the engine. The fuel tank, being lighter than the oxidiser tank, is set farther outboard than the other tank, for example, and this accounts partly for the 'unbalanced' appearance. Attitude control is otherwise effected through the reaction control system, which is very similar to that described for the Service Module. There are four 'quads' containing sixteen thrust chambers able to be fired in a pulsed or continuous mode and radiation cooled. Each chamber gives a thrust of about 100 lbf. Two parallel independent systems are provided, and there is a possibility of interchanging propellents between the various systems, since all propellents used are the same–50% hydrazine and 50% unsymmetrical dimethylhydrazine with nitrogen tetroxide.

The ascent stage houses a section which is essentially a complete spacecraft in its own right. It has an environmental control system, which provides a temperature- and pressure-controlled oxygen atmosphere in the Lunar Module cabin. This system also has to provide services to the two astronauts' pressure suits as well as to ensure a supply of water and oxygen for the portable life-support systems. Drinking water also has to be supplied and temperature control is necessary for all the onboard electronic equipment.

The general configuration of the ascent stage is seen in Fig. 14. The cabin differs from that of the Command Module in that the crew members stand in the course of the descent and ascent, and they do not need to have (neither is there enough weight allowance for!) the comprehensive couches provided in the Command Module.

It will be noted that there is an egress hatch near the bottom for descent to the lunar surface, and a docking hatch at the top, which joins on to the 'tunnel' at the forward end of the Command Module during crew transference. In addition to the main controls inside the crew cabin, and the various tanks and other pieces of equipment, a complex array of radio and radar aerials can be seen. These are associated not only with communication between the Lunar and Command Modules but include also the radar necessary for the final rendezvous between the ascent stage and the Command and Service Modules when the landing mission is completed.

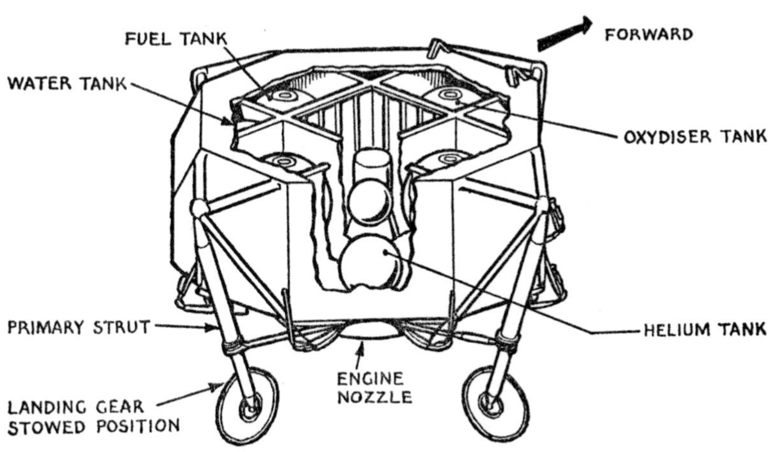

Fig. 16. Cut-away view of Lunar Module Descent Stage (simplified).

The Lunar Module cabin is 92 in in diameter and is made of welded aluminium alloy with a 3-in layer of insulating material and an outer cover of aluminium foil. The crew compartment is pressurised to 5 lbf/in^2 with a pure oxygen atmosphere and temperature is maintained nominally at 75° F.

Provision is made for sleeping, eating and waste management and full electrical power supply is, of course, provided both in the ascent and descent stages.

On the top of the descent stage is a thermal shield. This prevents the descent stage from blowing up–which could have a disastrous effect on the ascent stage–when the ascent stage engine is ignited during lunar take-off.

The Lunar Module also contains storage space for the two

experiments scheduled for the first lunar-landing mission. The first of these experiments is the detection of various characteristics of the lunar surface and a seismograph is built into this equipment. It will be able to transmit information back to Earth for a considerable time after the astronauts have left the Moon. The second experiment is a laser reflector which will be set up so that it can reflect laser signals generated from Earth. This will be used by many

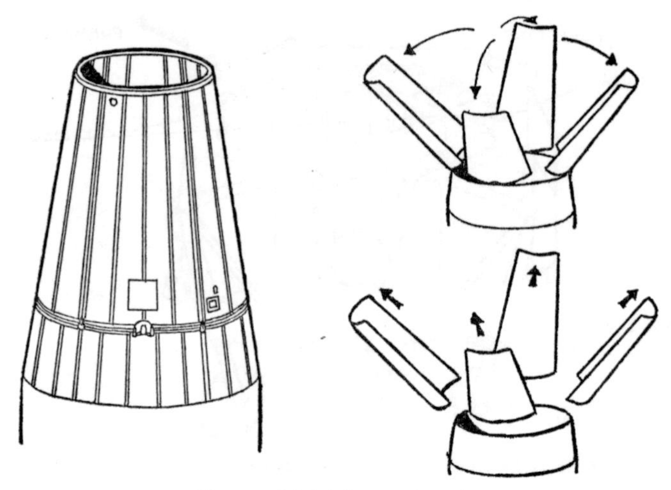

Fig. 17. LM-Adapter Panels.

scientists for a number of experiments, including some for determining more precisely the distance between the Earth and Moon at different parts of the lunar orbit. Additional seismographs and a range of scientific equipment will be ferried up in this storage space over the series of lunar missions in successive Lunar Modules.

We can see, then, that the Lunar Module is a further key element in the Apollo project; if it is not fully successful, then the whole programme fails in its aim to place men on the Moon. It can be seen that while the 'semi-direct' approach to the Moon involving the separation of a Lunar Module from the rest of the spacecraft in lunar orbit minimises the total size and energy required of the Earth launch-rocket vehicle, it introduces considerable complexity in having to provide a second self-contained spacecraft for the final descent to the lunar surface.

This section would not be complete without mention of the LM Adapter. This is an aluminium structure which protects the Lunar Module during launch and provides the structural attachment of the spacecraft modules to the Saturn V launch vehicle. The adapter is 28 ft long, and 12 ft 10 in diameter at the top where it is attached to the Service Module, and 21 ft 8 in diameter at the bottom where it meets the third stage of the Saturn V. It weighs just over 4000 lb and is built in four sections which, at the appropriate time, are released and opened by explosive charges. The separation is initially at the Adapter/Service Module interface, and the sections swing outwards like flower petals about hinges near the Adapter base. However, when the sections have swung out through 45 degrees, the hinges disengage and the sections are thrown out into space, this action being assisted by spring thrusters fitted near the hinges.

This operation both exposes the Lunar Module and frees the Command and Service Modules which, linked together, turn through 180 degrees, dock with the Lunar Module and remove it from the Saturn V third stage, after which the three spacecraft modules continue on their way to the Moon. The crew carry out this operation just after the trans-lunar injection phase from Earth orbit.

The Command and Service Modules have been produced by the North American Rockwell Corporation Space Division and the total contract is expected to reach the figure of $3300 million by 1970–about one-sixth of the total Project Apollo cost. Complex as these two modules may be, at first sight it seems quite unbelievable that such a large sum of money could be spent on them–where did it all go?

All straightforward design work–such as the design of a bridge or next year's car model–is a direct extrapolation of previous designs. When one tries to create something which has no comparable precedent, feasibility studies soon begin to show up large areas of ignorance, and this lack of knowledge can only be made good by creating mock-ups, carrying out experiments and building prototypes for testing. Rather than being a straightforward design operation it is more akin to finding out and learning on a vast scale.

The Project Mercury exercise had already shown the kind of problems that would have to be faced and the scale of development work required. For the Command and Service Modules–the former more particularly–the contract called first for twenty-three full-scale mock-ups. These were used by different groups within North American Corporation, by sub-contractors, and by NASA groups at the Marshall Spaceflight Center and the Kennedy Space Center, to study interior and exterior arrangements, handling and transportation, airlock and docking arrangements, lighting, systems integration, manufacturing and inspection methods–in many cases to build up virtually a new technology.

After this, thirty engineering test vehicles, termed 'boilerplate' models by the engineers, were scheduled. In the case of the Command Module these were complete modules structurally, but they carried ballast instead of all the complex internal systems. They were used for land and water impact tests, parachute recovery tests, and vibration and dynamic tests. Boilerplate models were also used to test all the various Abort Modes upon which crew safety depends. For the Pad Mode–when the Command Module would be blown off the Saturn V before lift-off in an emergency–a boilerplate module was blasted off from ground level by its Launch Escape Tower system and brought down by parachute. Similar tests were carried out at high altitudes and at transonic and high supersonic speeds in the atmosphere, it being remembered that the first stage of the Saturn V vehicle accelerates the spacecraft to well over 5000 miles/hour. Above 295,000 ft, when the Launch Escape Tower is jettisoned, to terminate the mission the Command Module has to be blasted off the booster rocket vehicle by the Service Module propulsion engine. In every case Modules were actually tested by launching, recovering and thoroughly examining them afterwards.

On top of this, the contract called for forty-nine manned or test Command and Service Modules, some for ground tests, and others for unmanned and Earth orbiting manned flights, leaving a residue of operational craft to carry out a series of lunar flypasts and landings. Finally, the contract required five test fixtures, four Apollo mission simulators, three evaluators, five trainers for crewmen, appropriate LM Adapters, and the necessary tracking and ground support equipment. In the event, some of the contracted vehicles were not required, and rarely it proved possible to refurbish a module to be used for another test–but, by and large, all 100-odd major items were produced.

The Lunar Module situation is very similar. The work given to the Grumman Aircraft Engineering Corporation will well exceed the value of $1500 million. In this case a team of around 7000 scientists, engineers and technicians have striven to produce the lightest possible vehicle consistent with the mission and the dependability required. Working on the very edge of what is currently practicable, they really have been stretched and, once again, there were no precedents or operational experience to fall back on. The original contract consequently was for two mission simulators and ten ground test articles as well as for fifteen Lunar Modules proper, that is flight models for final tests and operational use.

Unmanned flights do not usually make news, but a 'test article' Lunar Module was sent aloft in the Apollo 4 flight as early as 9th November 1967, this being instrumented to measure vibration, acoustics and structural integrity. The first unmanned test flight of the Lunar Module took place on 22nd January 1968, when the module's ascent and descent propulsion systems, along with the reaction control system, were successfully fired for the first time in space. The first manned Lunar Module carried out its separation, free flight and docking tests in Earth orbit over the first week of March 1969 in the Apollo 9 mission, which will be dealt with later.

A great deal of planning is necessary in this type of work, because in most cases new equipment, such as these modules, is flight-tested on booster rocket vehicles which are themselves in part or as a whole being tested in a certain stage of development.

These brief notes are, of course, quite inadequate in themselves to demonstrate the tremendous amount of development work which needed to be carried out before any manned tests were even contemplated–the mass of the iceberg, as one might say, underlying the tip that shows–but they can point the way if only by demonstrating how much hardware has, in fact, been built.

While the American astronauts are undoubtedly most gallant men, it should be appreciated that their lives and limbs are not lightly hazarded. Everything that can be tested and proved reliable for their missions is so tested and proved–and, of course, it is this painstaking thoroughness which costs so much.

1A Early Mercury Spacecraft being mounted on a Little Joe solid propellent rocket for test flight. In 1959 a rhesus monkey was recovered alive and well after the Escape Tower rocket was deliberately fired during a Little Joe flight. It is of interest to compare this photograph with that of the Apollo Spacecraft being lifted on to its Saturn V booster, only ten years separating the two pictures (see Pl. 9).

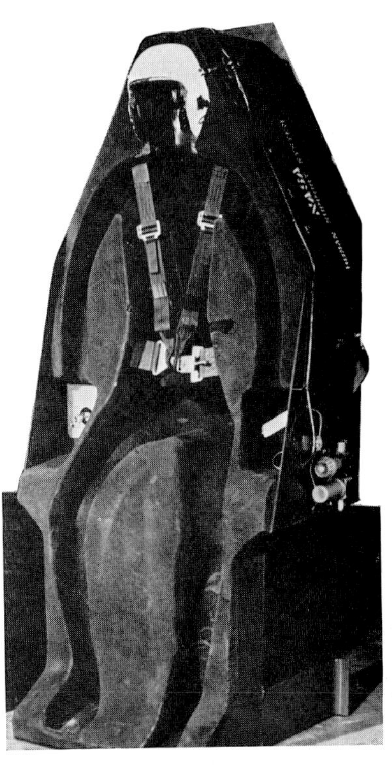

1B For the Mercury Spacecraft, where accelerations at launch and re-entry could be high, special one-piece seats like this were built and tested. Each seat was moulded to suit an individual astronaut and was marked with his name. It is interesting to compare this with the couches used in the Apollo Command Module, where accelerations are better controlled (see Pl. 5A).

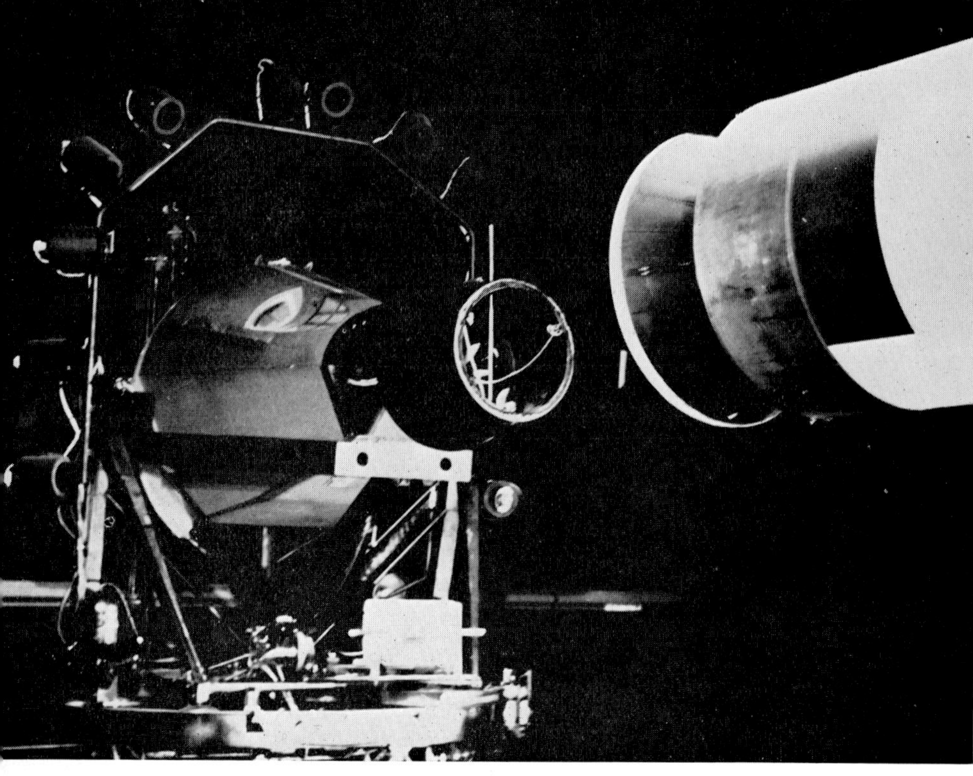

2A The Gemini/Agena Translation and Docking Simulator, part of the extensive astronaut training facilities at the Marshall Space Flight Center. The simulator was based on one used for docking research at the Langley Research Center.

2B Gemini 7 Spacecraft as photographed in Earth orbit from Gemini 6 after the two crafts' rendezvous on 15th December 1965, 160 miles up.

3A Astronaut Frank Borman is seen climbing a ladder in 1966 while five-sixths of his weight is supported by overhead straps. Wearing an Apollo pressure suit, he is simulating movements on the Moon, which has only one-sixth as much gravity as the Earth.

3B NASA Langley Research Center technician preparing one of several possible Lunar Mission models for tests to determine their Earth-landing characteristics. The models are catapulted by an overhead device, some landings being made in the water, others on the wooden platform. These tests (1961) helped in determining the final shape of the Apollo Command Module.

4A Fire belches with an ear-splitting roar from an F-1 rocket engine on a test stand in the Mojave Desert, California. A number of different test stands have been built, one of them capable of taking the entire cluster of five engines of the Saturn V first stage.

4B First stage of an early Saturn I launch vehicle being constructed at the Marshall Space Flight Center in 1959.

5A A view through the main hatch of the Command Module interior showing the three couches and main instrument panels. The layout of equipment and instruments changed somewhat over successive models.

5B View of the interior of the flame-scorched Apollo Command Module in which three US astronauts lost their lives in a flash fire on 27th January 1967. They were training for the first manned flight in the Apollo programme.

6A An Apollo Command Module and Service Module being assembled for extensive tests at Kennedy Space Center. (Photo: North American Rockwell Corp.)

6B The Apollo 8 Command Module being stripped down for examination after its successful flight round the Moon in December 1968. Note the charred base and paintwork as a result of re-entry heat. The tunnel for entry to the Lunar Module can be seen at the top and round it the space for parachutes and other Earth landing equipment. (Photo: North American Rockwell Corp.)

7A One of a small family of aircraft with rebuilt fuselages to carry heavy Apollo programme equipment. This plane, christened 'Pregnant Guppy' because of its shape, breaks in two for loading and unloading its cargo–in this picture a Saturn S-IV rocket stage.

7B The Crawler-transporter is here seen being positioned beneath a Mobile Launcher prior to moving it into the Vehicle Assembly Building.

8A The Crawler-transporter is seen here moving a Mobile Launcher into the Vehicle Assembly Building. The Launch Control Center is the low building to the left.

8B A view from ground level of the three Saturn V rocket stages assembled on a Mobile Launcher in the Vehicle Assembly Building.

9 The Apollo 10 Spacecraft is here seen being lowered for assembling with the Saturn V rocket's instrument unit. Technicians can be seen inside the instrument unit waiting to make connections. This photo was taken from the top of the Mobile Launcher tower, and the 'White Room' at the end of its swing arm can be seen to the left.

10 A night view of the Saturn V vehicle waiting on the Launch Pad. Taken from the top of the Mobile Launcher tower, this picture shows clearly the Astronauts' Access Arm and the various Service Swing Arms.

11 The Crawler-transporter is here conveying the Mobile Launcher with its Saturn V load up the ramp of the Launch Pad hardstand. Notice how the hydraulic cylinders on the transporter have kept the Launcher and Saturn vehicle vertical. Notice also the small group of men standing at the left corner of the Launcher deck, which demonstrates well the huge size of the structure.

12 The first Saturn V lift-off on 8th November 1967.

13A Here the Apollo 10 Lunar Module (LM 4) has been fitted with the bottom part of the Spacecraft LM Adapter at the Kennedy Space Center and the top part is about to be lowered for fixing.

13B Apollo 11 astronauts Neil Armstrong left and Edwin Aldrin in the Lunar Module simulator practise their descent to the Moon's surface.

14A Apollo 10 Command Module 'Charlie Brown' photographed 60 miles above the Moon's surface from the Lunar Module 'Snoopy'. This and the next photograph were taken on the far side of the Moon where the craters are unnamed.

14B Apollo 10 Lunar Module 'Snoopy' minutes before docking with the Command Module. In the lower left corner can be seen the rendezvous radar dial. One of the far groups of thrusters can be seen clearly.

15 Aldrin leaving the Lunar Module hatch to follow his companion on to the lunar surface. This photograph was taken by artificial light, as this part of the spacecraft was in deep shadow. Note the slimness of the ladder rungs, which would break under Earth gravity conditions, and the crinkled aluminium foil round the descent stage for reflecting the sun's heat.

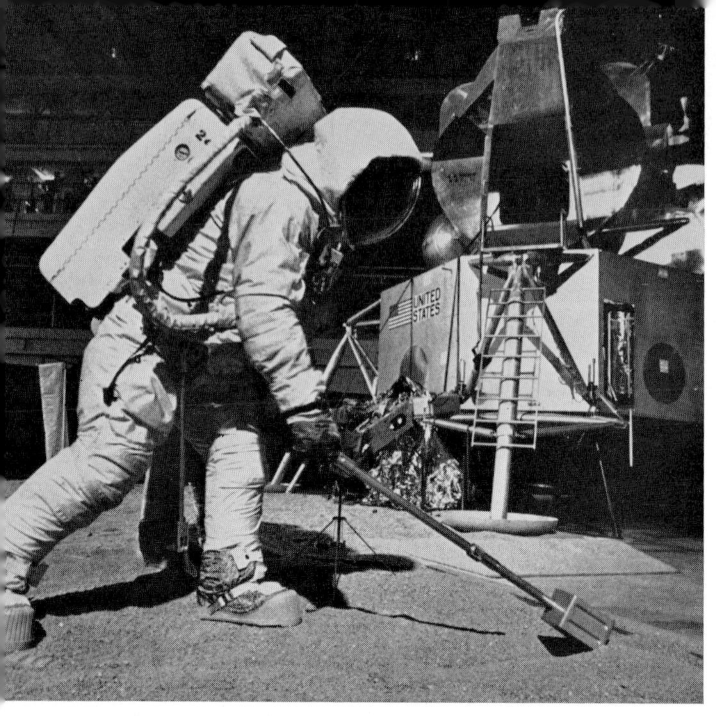

16A Edwin Aldrin practising the various lunar extravehicular activities on Earth with a Lunar Module simulator behind.

16B Edwin Aldrin deploying the Passive Seismic Experiments Package on the lunar surface on the Apollo 11 mission, 21st July 1969. By his right hand is the Laser Reflection Experiment. The small white post on the horizon is the TV camera mount.

8

The Astronauts and Life Support

SINCE we are discussing the American-*manned* space programme, it is time something was said about the astronauts themselves and their spacecraft environment.

Between April 1959 and August 1967, NASA made six selections of personnel for training as astronauts, a total of sixty-six being chosen out of 3000 odd applicants. Of these seventeen were taken as scientists and will be used in later more scientific missions. The other forty-nine were basically in the 'test pilot' category. Indeed, the initial three criteria which had to be satisfied were that applicants had to be US citizens, no taller than 6 ft and no older than thirty-five at the time of application; they had to have a Bachelor's degree in engineering or science; and they had to have acquired 1000 hours jet-pilot time or have graduated from an American Armed Forces Test Pilot School. Even the scientist intake had to learn to fly after acceptance, at least one being eventually dropped because he was unsuccessful in this line–and four astronauts have died in air crashes.

Astronauts have to be superbly fit, but this does not mean just physical fitness. A kind of psychological fitness is one of the prime qualities looked for among candidates. How they react 'under pressure' and in serious situations is important. In initial fitness tests carried out at the Aerospace Medical Laboratory, candidates were put into heat chambers and subjected to a temperature of 132° F (56° C) for 2 hours. They were put into chambers where the air pressure was reduced to that at 65,000 ft and sometimes more. Those who came through these and other tests were judged to be extremely well-adjusted individuals. They had to be people capable of working with others as a team, men with courage and resourcefulness, 'men in their physical prime, young enough to be resilient physically, and yet mature enough to have lost the rash impulses of youth', as James Webb, until recently Administrator of NASA, put it.

The first gallant group selected in 1959 were faced with a great unknown–no one had yet been into space and all was speculation. The Project Mercury astronauts not only had a great deal of theoretical studies to cope with in the classroom –they were shut in darkened cells for considerable periods of time to experience complete isolation; they were taken up in aircraft which, for a short time, flew on a parabolic course designed to match the flight path of a missile, so that they were in the 'free fall' condition and could experience what weightlessness was like; within their pressure suits they lay submerged in water to simulate weightlessness over long periods; and they were spun round in a centrifuge to experience the effect of high acceleration (*g*) forces, when they 'weighed' many times their own weight. They also worked on the actual flight hardware and made 'flights' in simulated capsules.

Much of this early training was necessarily experimental in that everyone, teachers as well as taught, were learning as they went, and throughout there has been an inevitable interaction between astronaut training and hardware testing. Astronauts not only learnt from handling and operating spacecraft and equipment but they were also able to report back information which led to better and more easily used hardware. Nor was this training and testing confined to land. American spaceships, from the start, were designed to come down in the oceans of the world and the crews had to be prepared for the immediate hazards of landing on water and for survival at sea in case there was a delay in their being picked up. Astronauts have spent hundreds of hours of training getting out of spacecraft, bobbing around both in tanks and in real ocean environments in all weathers, and experimenting with survival kits and rubber dinghies. Indeed, this is one part that many like least of all!

Besides keeping well up to date with space-flight techniques, the astronauts chosen for any particular mission have to carry out extensive training programmes relevant to that flight. The crew members for the historic Apollo 8 flight round the Moon, for instance, spent more than seven hours of formal training for each hour of the mission's 6-day duration. About 1100 hours of training were scheduled for this crew over and above normal preparations for the mission.

It would be tedious to list all these areas of training but, apart from the more obvious aspects, such as fire and emergency precautions, practice at evacuating the spacecraft in the Gulf of Mexico and helicopter pick-up–all these things

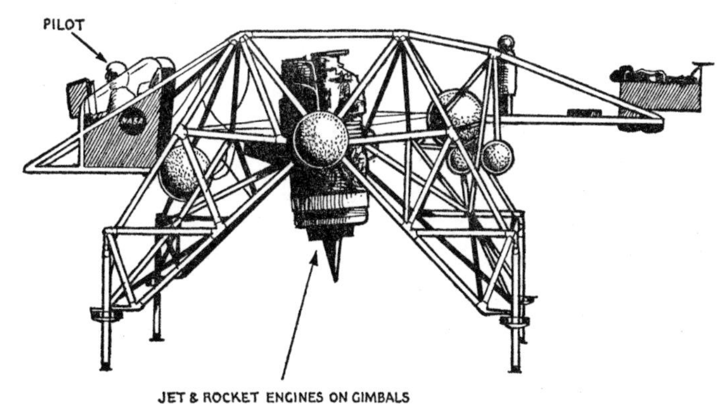

Fig. 18. Lunar Module Landing Simulator. Later versions were made for astronaut training.

are rehearsed–crew members spent some time in a planetarium so that they could study the stars for navigation purposes. They also had detailed briefings on the spacecraft systems, operation and modifications, participating in the spacecraft manufacturing check-outs and becoming generally conversant with all the complicated systems which would make their trip possible. And there have usually been scientific experiments of one kind or another on each voyage, which had to be practised.

Similar training schedules, but appropriate to individual missions, applied to all crews. For Project Gemini, the Langley Research Center constructed a special rig to examine the problems of space docking with another vehicle. This was, in fact, a vehicle-docking simulator, and a similar one for training purposes was derived from this and used extensively by the astronauts concerned.

One of the most difficult areas of training has been that concerned with the Lunar Module. This, being meant for use in the lunar environment, is a rocket-sustained flying machine with its attitude controlled by small jet thrusters. It cannot be flown in the Earth's atmosphere because our gravity is six times that of the Moon–it would weigh six times as much and need rocket motors six times as powerful, and they would not work as efficiently as they would in a space vacuum.

But simulators were needed both for training purposes and for gaining practical data for the Lunar Module design. The first simulator constructed was a peculiar vehicle made up from a helicopter cockpit, several available rockets and a tubular framework. It was supported in a special gantry at the Langley Research Center. This was 400 ft long by 260 ft high and the craft was supported by a cable system which took five-sixths of its weight, the rockets being used to lift the other one-sixth. This rig allowed a better understanding of the problems relating to a Moon landing.

A later simulator provided greater realism and more flexibility. It was a specially designed machine which bore little resemblance to the actual Lunar Module, except that it had four legs. It had two lift systems. One was a jet engine supplying a constant downwards thrust equal to five-sixths of the machine's weight. The other was a variable thrust-rocket group similar in operation to the thrust rockets of the Lunar Module. Depending upon whether this rocket thrust was more than, equal to or less than one-sixth of the vehicle weight, the machine tended to climb, hover or descend. The original machines, called Lunar Landing Research Vehicles, were delivered in 1964 and were used essentially for 'finding out', but two were later delivered to the Manned Spacecraft Center for astronaut training, and three more machines of an improved type, called Lunar Landing Training Vehicles, were delivered later. There have been some accidents with these craft, but they seem largely to have been caused by wind shear and other atmospheric conditions which are absent on the Moon. All the same, flying a craft like this is still a tricky business.

The Lunar Module, of course, leaves the Command Module at a height of some 80 miles above the lunar surface, although this planned height could be varied somewhat in the actual missions. In any event, for practice purposes a complete landing cannot be made from that height on Earth. Consequently, a Lunar Landing Approach Simulator was built around a closed circuit television system, very similar to those used in aircraft simulators for pilots. The full-scale model of the spacecraft is suspended in the centre of a large sphere, projection television being used to give a picture of the part of the Moon being overflown on the sphere's inner surface. Outside, and unseen of course by the crew, is a large model Moon photographed continuously by a television camera running on rails and able to move left or right, up or down, and in or out, these movements being those appropriate to the flight-control signals put in by the spacecraft pilot, who sees the lunar surface from his craft just as he would in the real situation. Because the range covered is large–from a few hundred miles above the Moon's surface right down to virtually landing–in fact four models of the lunar surface were constructed, each being on a larger scale than the preceding one. Only one is a complete globe; the others are just parts of spheres, the last being of such a large scale that it is flat. As a camera pans in towards its limits on one model, the camera on the next larger scale model takes over and so on, the process being reversed for leaving the lunar surface.

Similar techniques, using an Earth model and closed-circuit television, were used to accustom astronauts to seeing the Earth receding from them, so that they would be more able to accept this experience on an actual voyage to the Moon, and so that they could make practice navigational sightings.

Unlike the astronauts, the rest of us are completely Earth-bound, and most of us take our natural living environment so much for granted that we underestimate the problems of providing a small local environment in which a man can function in safety and reasonable comfort.

A man needs something like 2½ lb of food, 4–5 lb of water and nearly 3 lb of oxygen a day at normal temperatures to meet the needs of a working day with an energy output of 500 BThU/hour (176 cals/hour). On the Apollo voyages in space, the average daily value of three meals is 2,500 calories per man, food being either 'bite size' like a biscuit and complete, or dehydrated in plastic containers. In the latter case the food is rehydrated by injecting water in 1-oz increments from a spigot at the food preparation station, one tap supplying cold water at 55° F (13° C) and the other hot water at 155° F (68° C). There is a dispenser for drinking water which emits ½-oz spurts at each squeeze. These arrangements are necessary because in the weightless condition water will not pour. The water is supplied from the Service Module fuel cell by-product water.

On Earth we breathe a complex gas mixture, roughly 20% oxygen and 80% nitrogen, together with small amounts of argon, carbon dioxide, and varying amounts of water vapour. Atmospheric pressure at sea level is 14·7 lbf/in^2 and the partial pressure of oxygen, i.e. the pressure the oxygen would exert if the other gases were not present, is about 3·7 lbf/in^2. In other words, if one lived in an atmosphere of pure oxygen, one would need a pressure of 3·7 lbf/in^2 to exist as at sea level. With acclimatisation, a man can manage with an oxygen pressure down to 2·7 lbf/in^2, but below this lies severe hypoxia, leading generally to collapse.

The lungs not only take in oxygen but they also give out carbon dioxide, and how well they can discard this impurity depends upon the partial pressure of carbon dioxide in the atmosphere being breathed. The normal sea level partial pressure of carbon dioxide is only 0·004 lbf/in^2 compared with the total air pressure of 14·7 lbf/in^2, and the permissible level is very much less than most persons suppose, the maximum allowable concentration of carbon dioxide for long term working being only about 0·1 lbf/in^2. Beyond 0·3 lbf/in^2 partial pressure lies carbon dioxide narcosis.

There is also an output of water from the lungs and the skin, the amounts depending upon the body temperature, the air temperature, the humidity and so on. Consequently, it will be apparent that the environmental control system in a spacecraft is tremendously important, for although man can withstand quite abnormal conditions for short periods of time, the range of abnormality is fairly small for long periods. The temperature and the carbon dioxide and humidity conditions must be continuously monitored and kept within close limits to the ideal. Temperature comes into the picture because men generate metabolic heat as a result of their body activities and if this is not removed they overheat, sweat and generally become very uncomfortable–and even if this moisture is removed, the body would still be doing a lot of hard work to keep cool, and the environmental control system would be working overtime.

When the Americans were designing their first spacecraft various decisions had to be made, some of which have had and will have long term effects.

Firstly, nobody knew enough about space conditions to consider putting astronauts into a sealed, pressurised compartment in 'shirt-sleeve' conditions. What would the leak-tightness and general integrity of the spacecraft structure be? What would happen if a meteoroid penetrated the shell so that the cabin was evacuated of air? To play safe, the designers accepted the idea that the spacecraft crew would be dressed in pressure suits so that in the event of a cabin-pressure failure, the crew would be safe.

They might not be so comfortable, though, because while in normal conditions the gas pressure would be the same in the cabin as in the pressure suit, with the cabin evacuated of air, the suits would be 'blown up' by the internal pressure, which would make it difficult for the crew members to move. Of course, all kinds of developments could be looked forward to, and a pressure suit could be anything from a strong enough airtight fabric covering to a complete 'suit of armour'.

Right from the start, it was obvious that to get anywhere, spacemen needed to be able to operate in as normal conditions as possible, and if a suit had to be worn it should be a lightweight one. This meant keeping the gas pressure low, and NASA settled on 5 lbf/in^2 pure oxygen atmosphere for cabin and suit as basic. In the event of a cabin depressurisation, the suit pressure would drop to 3·7 lbf/in^2–equal to the partial pressure of oxygen at sea level.

Only a year or two in advance of Project Mercury was the North American X-15 rocket plane, which started flying in 1960 and made a number of flights out of the Earth's atmosphere, up to 67 miles for short periods. Cabin pressure was regulated at 3·5 lbf/in^2 above that outside the cockpit, but two gas systems were used. Pure oxygen was breathed by the pilot in his pressure suit, while the cabin was pressurised with nitrogen to avoid fire risks, which were considered would be too great if the cabin were filled with pure oxygen, especially at lower levels in the atmosphere when pressure rose.

When it came to the Project Mercury space capsules, it was decided to use only one gas–pure oxygen–at 5 lbf/in^2 pressure, the environmental control system having effectively two loops, one through the cabin and one through the pressure suit. Firstly, this saved the weight of having two gas systems and equipment. Secondly, the conditions were different from the X-15; there was not so great a fire hazard aboard the Mercury craft as in the X-15; and in the long run it was hoped at least to experiment with 'shirt-sleeve' conditions, in which case the cabin atmosphere had to be the same as that in the suits. All the same, the original environmental control system for the Apollo Command Module, though more complex than Mercury's, still envisaged the astronauts in pressure suits, their air and the cabin air being processed by the same system.

The environmental control system for the astronauts consists essentially of a circuit containing duplicated fans or compressors to drive the air round through carbon dioxide and odour absorbers, and a heat exchanger for cooling the air and condensing out water vapour. This condensed vapour becomes waste water, and the evaporation of this at the vacuum pressures outside the spacecraft provides the cooling mechanism in the heat exchanger. Water, expanded to steam under vacuum conditions, has a heat extraction potential of 1030 BThU/lb. In the spacecraft as a whole there are other generators of heat, such as electrical equipment, and these are cooled by another system using a cabin heat exchanger and a glycol circuit which takes waste heat to space radiators in the Service Module. The systems are interconnected, however, to increase reliability.

The temperature in the Apollo Command Module is kept at 75° F (24° C) and oxygen used up is replenished from two oxygen supply tanks in the Service Module. There is also a small oxygen tank in the Command Module for the re-entry phase, when the Service Module has been jettisoned.

During the Gemini flights more confidence was gained in spacecraft integrity and micrometeoroids were not found to be as bad a menace as some had feared. Experiments were therefore made with working in 'shirt-sleeve' conditions, and the results were so successful that modifications were made to the Apollo spacecraft and the Apollo astronauts now operate in these conditions all the time, except for the critical conditions applying at lift-off, lunar landing and so on.

In the spacecraft now, the crew have on a constant wear garment over which is worn either the in-flight coveralls or the pressure suit. The constant wear garment is a one-piece suit similar to 'long-johns' and is of porous-knit cotton with a waist to neck zipper. In the 'shirt-sleeves' condition the crew wear teflon fabric in-flight coveralls.

In more critical operations they don what are called Intra-vehicular Pressure Garment assemblies–'space suits' to most people. They consist of a multi-layer suit, with helmet and gloves which can be pressurised independently of the spacecraft. The outer suit layer is teflon-coated beta fabric woven of fibreglass strands; under this is a restraint layer of Dacron and Teflon link-net to restrain internal pressure and maintain the suit's shape; beneath this is a black neoprene-coated nylon garment referred to as the pressure bladder; and finally, there is an inner high temperature nylon liner, which replaces an earlier simple comfort layer. Oxygen connections, communications and biomedical data lines are attached to fittings on the front of the torso. With the pressure helmet on, communication carriers are provided inside, doubled up to give two microphones and two earphones for reliability.

The helmets have varied over time and for the various

types of mission, but the basic pressure helmet attached to the pressure bladder is of the 'goldfish bowl' type, while for extra vehicular activity a metal over-helmet is fitted containing a glazed visor. The pressure suits have been developed over the years as the result of experience and new materials and techniques, and now weigh about 60 lb.

For the 'walking in space' operations, modified suits known

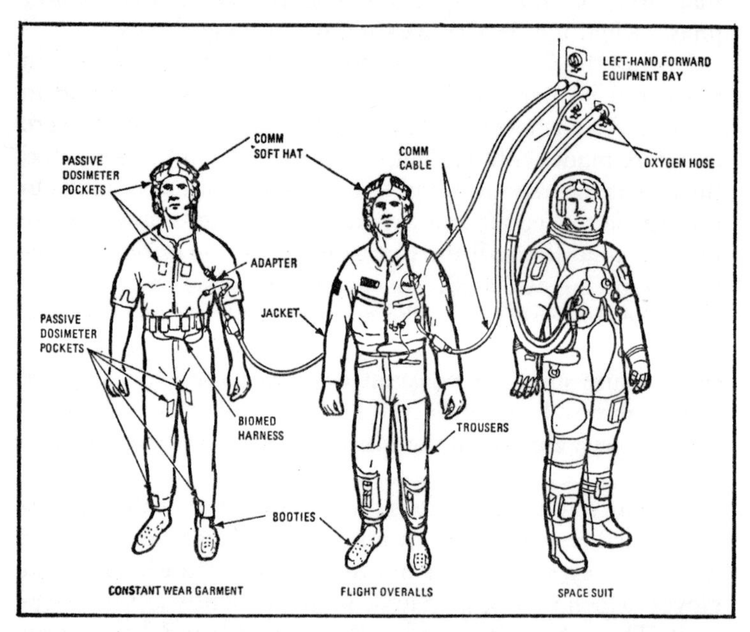

Fig. 19. Constant Wear Garment, Coveralls and Spacesuit for use inside the Spacecraft. A Water-cooled Undergarment and an outer suit covering are donned for extravehicular activity.

as Extravehicular Pressure Garment assemblies have been used. These were the basic Intravehicular suits with an additional outer covering having up to twenty-one layers of material able to ward off the micro-meteoroids encountered in space, and the helmets had a specially coated visor which only passed 10% of the light to save an astronaut from eye injury in case he looked towards the sun.

The astronauts in the Gemini craft donned their Extravehicular suits, evacuated the cabin atmosphere and opened the hatch. The space walker then left the vehicle and floated weightlessly outside. Of course, for safety he was tethered by a rope. He also had an umbilical hose for oxygen attached to the spacecraft environmental-control system. He could have existed in space just on this; however, these extravehicular activities were not just playing around; they were part of tests needed to develop a suit for use, say, on the lunar surface or on other activities in which an umbilical attachment would be inconvenient, if not impossible. Consequently, these astronauts carried an Extravehicular Life-support System and later ones also carried on their backs an Extravehicular Support Package. The two units were connected together to form virtually one unit. Connections from the front unit to the space suit made basically a closed oxygen circuit; however, this atmosphere was circulated round the suit and circuit by oxygen fed through an injector-type pump, this oxygen coming either from the oxygen cylinders carried by the astronaut, or from the spacecraft through the umbilical line. Overflow from the circuit was dumped into space. Since fresh oxygen was being added all the time in this system, no provision was made for carbon dioxide extraction, but there was an evaporator/condenser for cooling and water vapour extraction.

The pack on the space walker's back contained a freon cylinder, which was connected through valves to a Hand-held Manoeuvring Unit–a portable jet-thruster which, with practice, allowed an astronaut to position and orientate himself in space, much as the spacecraft themselves were controlled. The back pack also contained an ultra-high-frequency voice transceiver and aerial antenna, connected to the space suit and to the spacecraft through umbilical leads, which also carried other information, such as signals from pressure transducers registering oxygen and freon supplies.

Space-suit equipment has, naturally, been continually under development and while the suits for the lunar landing are basically of the earlier extravehicular type, there are differences in the overall equipment. Apart from many minor changes and additions, such as lunar overshoes, the main differences lie in the undergarment and the Portable Life-support System. In place of the constant wear garment, the astronaut dons a Liquid Cooling Garment. This looks like underwear made from white net material. In fact, it is a nylon Spandex material supporting a network of Tygon tubing through which water from the Portable Life-support System is circulated.

This support system is carried as a back pack and, apart from the cooling liquid, it sustains oxygen, telecommunications and electrical systems. With this equipment oxygen is supplied at a temperature of about 45° F (7° C) and returned to the pack at about 80° F (27° C) carrying with it carbon dioxide, body odours and water vapour. It passes through a canister containing deactivated charcoal and lithium hydroxide, which removes the carbon dioxide and purifies the gas. The cooling water and the oxygen circuit are cooled in a heat exchanger in the pack relying on evaporative (sublimative) cooling. The life-support system will operate for 4 hours with a crewman output of 1600 BThU/hour and peak rates of up to 2000 BThU/hour.

The lesser lunar gravity led space fiction film-makers to portray astronauts on the Moon walking around almost effortlessly. Such is not the case; at least the indications to date suggest that this is not so. Training rigs have been used on Earth in which astronauts have had five-sixths of their weight carried on overhead support straps, and trainees have climbed steps and carried out various tasks in this condition. Walking depends not only on the leg muscles pushing the body upwards a few inches but also on gravity pulling it down again. With a lower gravity the step becomes much slower and walking–especially in a pressure suit and carrying a back pack, which even on the Moon will weigh 11 lb–proves very tiring. The best way, it was thought, to move about on the Moon's surface would be to make 'kangaroo hops'–but this has proved to be incorrect.

The amount of time the first astronauts spend on the Moon is considerably shorter than that originally planned. For all the great advances in technology made during and through the Apollo programme, when it comes to the actual landing, the space suits and other equipment used are still only marginally sufficient, although, considering the nature of the problems, it has really been quite an achievement to create suits and life-support equipment of reasonable proportions which will sustain men in the harsh vacuum conditions of space and allow them to function with so few restrictions.

Finally, astronauts being human, there is the question of the disposal of body wastes. The methods in use in the Apollo spacecraft were developed and tried out over the Project Gemini flights. Solid body wastes are collected in plastic bags with adhesive lips, which provide a secure attachment to the body. The bags contain germicide, which prevents the formation of bacteria and gas, and the adhesive lip is used for sealing the bag after use. At present bags are stored and brought back for contents analysis. Urine collection devices are provided for use either while wearing the pressure suit or while in coveralls, and the urine is dumped overboard through a spacecraft dump valve. This applies only to the Command Module. As it is the policy of the space powers not to contaminate other worlds, vehicles and equipment are sterilised before leaving Earth and precautions are taken so as not to contaminate the lunar surface.

In the small Apollo Command Module, astronauts cannot wash in the usual sense for personal hygiene, though experiments are being made with showers for use in Earth-orbiting stations. Consequently, each man-meal package contains a wet-wipe cleansing towel, and there are ample supplies of 12-in square dry towels and dispensers containing 3-ply tissues.

While the problem of sustaining human life in space for the short time span needed for a lunar mission has been solved satisfactorily, it will be apparent that a great deal of work will need to be done and many more facilities will have to be provided before man can be sustained physically, psychologically and emotionally for the many months of isolation which would be necessary to venture farther afield to the planets.

The way is certainly not yet wide open for long space missions. Over periods of time spacemen lose bone calcium in the weightless condition and unless a cure can be found or it is proved that this process stops at some threshold level, astronauts could pay a heavy price for their exploits. Again, while no harm comes from breathing pure oxygen over short periods, there are believed to be long-term harmful effects, which could lead to a changeover to an oxygen atmosphere diluted with nitrogen, helium or some other inert gas at a higher pressure, such as the Russians have used all along.

9
The Saturn Launch Vehicles

SOON after the beginning of the space era and the first satellites of 1957, consideration was given to the evolution of a family of rocket vehicles which could be used for the major space missions, including that of sending men to the Moon. The original rockets such as the Jupiter and Thor, were based on a rigid tank construction in which light alloy was formed into tanks to contain the propellents and oxidisers, much in the form of the then aircraft construction techniques. Although good for their time, these rockets only had a medium structural efficiency, as the rocket-engine thrust was carried up through the tank walls to the upper stages and payload. This was better than in the early V2 rockets, which were formed of a rigid structural framework with an outer cover largely for aerodynamic reasons only.

Greater structural efficiency was attained in the Atlas type of construction, in which the tank wall thickness was reduced to some $\frac{19}{1000}$ in, and the material changed to high-tensile stainless steel. Structural integrity was maintained by internal pressure in the tanks, rather like the rigidity given to a sausage balloon when it is blown up. The engine thrust was then transmitted through the fuel in the tanks and the pressurising gases. The same principle is used for the British Blue Streak vehicle, which forms part of the European Launching Development Organisation's rocket, Europa I and II, and is used in later American rocket vehicles.

However, the early Atlas vehicle and even the Titan II, which were used for the Mercury and Gemini programmes respectively, could only put up payloads of a few tons in Earth orbit and for lunar and planetary shots, the payloads would be more in the region of a ton or under. It was pretty obvious that a larger intermediate vehicle was required to carry out certain tests, particularly with respect to the prototype development of the Apollo spacecraft and the Saturn I was conceived. This vehicle actually went through three stages of development. The original Saturn Is, called Block 1, had dummy upper stages and were just to prove the basic concept. The Block 2 versions had a live second stage designated S-IV which was notable in that it was the first time liquid hydrogen had been used as a fuel in a large space rocket. Finally, there was an up-rated version called the Saturn IB, which had a new second stage called the S-IVB. It is not intended to discuss all these vehicles here and only the Saturn IB will be described.

The Saturn IB was the development responsibility of the George C. Marshall Space Flight Center at Huntsville, Alabama, under the direction of Dr Wernher von Braun, who is perhaps one of America's best-known rocket-vehicle engineers. The vehicle was designed so far as possible on the basis of known technology and available equipment, rather in the same way that the Russians developed their Vostok and Voshkod launchers on the basis of technology built up on ballistic missile development. In both the US and the USSR cases, engines were clustered to provide the appropriate thrust. In the Saturn IB the tanks in the first stage were visibly clustered also, there being eight tanks of the Jupiter vehicle-type clustered round an inner core tank, leading to a diameter of 22 ft at the rear propulsion bay end.

In the propulsion bay itself, eight identical engines were used as in Fig. 20, each with 188,000 lbf thrust, giving a total thrust of over 1·5 million lb. With the second stage, this vehicle was able to place 20 tons into a low Earth orbit, which would allow it to test substantial parts of the Apollo spacecraft in Earth orbit. It was important, however, also in proving the second stage, the S-IVB, which was to be the third stage of the larger Saturn V vehicle. The Saturn IB, therefore, represents an element in the building block concept in the development of US launch vehicles, where stages and parts could be tested in one configuration and then used in another.

Fundamentally the Saturn IB consists of two stages and an instrument unit as in Fig. 20. The first stage, known as the S-1B is built by the Chrysler Corporation, while the second stage is produced by the McDonnell-Douglas Corporation. The first stage uses kerosene and liquid oxygen as fuel and oxidiser, whereas the second stage is a large all cryogenic booster using liquid hydrogen at minus 423° F and liquid oxygen at minus 297° F. The combined vehicle stands 138 ft tall with its two stages, but reaches 224 ft in height when carrying the Apollo spacecraft on top. The weight of the complete vehicle dry is 150,000 lb, but with propellents this

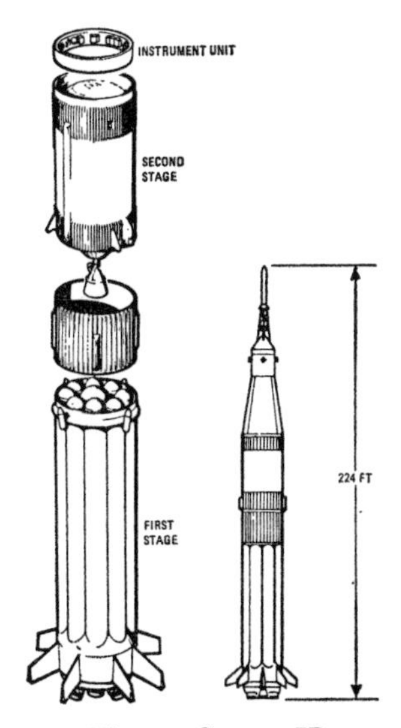

Fig. 20. Saturn IB.

reaches 1,300,000 lb. The S-IVB stage is 58 ft in height and has a diameter of 21 ft 8 in with a gross weight of 253,000 lb. It has one J-2 rocket engine burning liquid hydrogen and liquid oxygen which gives a thrust of 225,000 lbf in vacuum.

The time-scale for the development in the Saturn series of vehicles can be seen from the following programme. The first Saturn I with only a live first stage was launched successfully on 27th October 1961, and this was followed by three more Block 1 launches, which led on to the first Block 2 lift-off on 29th January 1964, this having a live second stage, nose cone and boilerplate spacecraft. Altogether, six Block 2 models were fired, three of them being used for work outside the Apollo programme, and the first Saturn IB, with the hydrogen burning second stage, flew on 26th February 1966.

The technology gained in developing these vehicles was

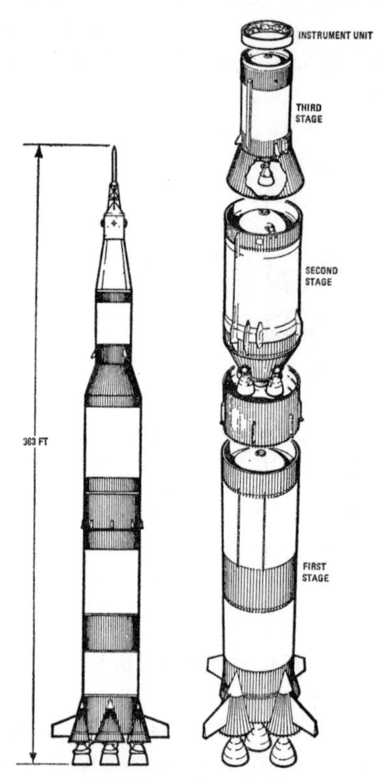

Fig. 21. Saturn V.

essential to the design of the much larger Saturn V. New tools, jigs and fixtures had to be designed and built and new manufacturing techniques had to be developed before actual fabrication of the Saturn V could begin. The assembly, check-out and countdown experiences with the Saturn I led to a new kind of launch pad and new assembly procedure as we shall see in another chapter. The vast size of the Saturn vehicles required also the construction of new, giant test facilities both at the Marshall Space Flight Center in Huntsville and at the Center's Mississippi Test Operations Site in Hancock County, Mississippi. A factory near New Orleans was reactivated and converted into an assembly plant for the large first stages, this being known as the Michoud Facility. And other new manufacturing and testing facilities had to be provided for the second and third stages and the powerful new J-2 and F-1 engines.

As its plans took shape, the Marshall Center took its needs to industry, negotiating contracts with dozens of major contractors, who in turn negotiated sub-contracts with thousands of other suppliers. Scientists and engineers in the several laboratories at the Marshall Center worked hard testing materials, developing new materials and revolutionary welding methods, making aerodynamic studies, and installing and programming the many computers set up to solve the myriad of engineering and other problems. Test monitoring and recording instruments in vast numbers were installed, checkout procedures developed, reliability standards established, electronics systems designed and built and guidance schemes developed.

Out of all this came the Saturn V. The first S-1C stage–the largest of those in the Saturn V vehicle–was rolled out of its Marshall Space Flight Center hangar on 1st March 1965. It was a static test model and it was test-fired in a stand for the first time on 9th April 1965. The first few of the S-1C stages were built by the Marshall Center, but subsequent first stages were provided by the Boeing Company from the Michoud Facility, at New Orleans.

The Saturn V has been an immense engineering operation. The three vehicle stages stand 282 ft high, and with the Apollo spacecraft on board the height is 363 ft. The vehicle proper weighs 430,000 lb dry, and 6·2 million lb fuelled up. Its payload capability into a low Earth orbit is 270,000 lb, over ten times that of Saturn I, and it is capable of inserting 100,000 lb into a trans-lunar trajectory–the maximum weight of the Apollo spacecraft.

A number of engines have been used in the various Saturn

vehicles' stages and these cannot be discussed here in detail. However, the general construction of these engines is worth noting. They all consist of three main parts–a combustion chamber into which the fuel and oxidiser are injected, a pump for pushing these fluids into the combustion chamber,

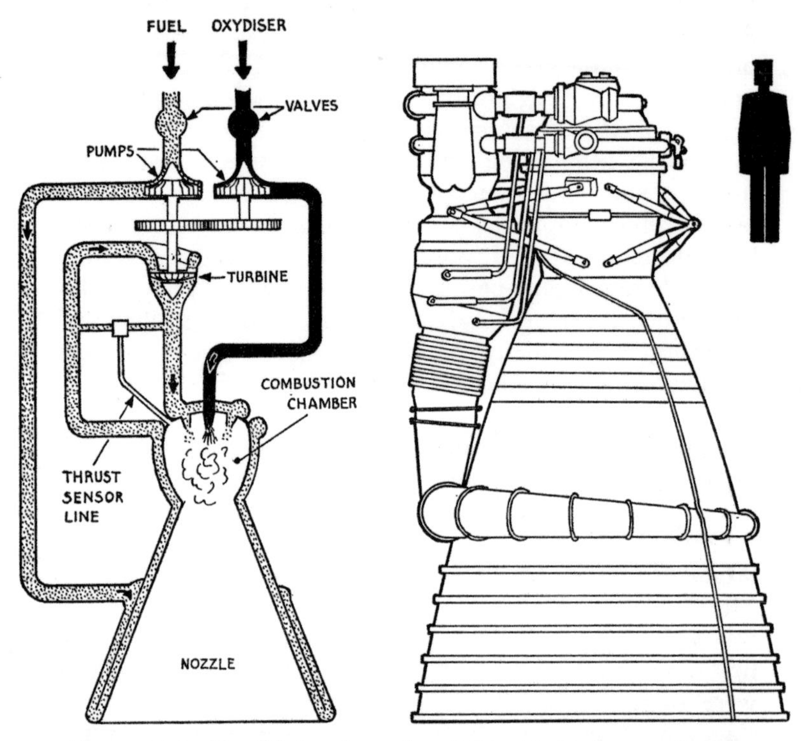

Fig. 22. *Left:* Simplified schematic of Liquid Hydrogen Rocket Engine. Rocket Engines vary in detail, some having a separate Gas Generator.
Right: The F-1 Engine used for the Saturn V S-IC stage with a man for comparison of size.

and an expansion nozzle which allows the hot gases to expand most efficiently from the combustion chamber. The nozzle and combustion chamber surfaces have to be cooled, otherwise they would just melt and burn up in the heat generated, and in most large liquid-fuelled rocket engines this is done by giving them hollow walls and circulating the fuel round these parts before injecting it into the combustion chamber for burning.

We can trace the main sequence of events through a typical hydrogen-burning engine, although they all, naturally, differ somewhat in detail. The hydrogen starts as a liquid at minus 423° F in the fuel tank. It passes through a fuel inlet valve, through the fuel pump, through the nozzle-cooling jacket where, by cooling the nozzle walls, its temperature rises and it becomes a gas at minus 100° F. This expands through a turbine furnishing the mechanical power to pump more hydrogen from the tank into the combustion chamber via the nozzle jacket and pump turbine. The turbine also pumps liquid oxygen from its tank directly into the combustion chamber where it burns with the gaseous hydrogen. To control the thrust there is a sensing line from the combustion chamber. If the thrust runs too high, some of the gaseous hydrogen is allowed to by-pass the turbine, so that the pumps slow down slightly and reduce the propellent flow rate, which in turn reduces the thrust.

The largest rocket engines used by the Americans are the F-1 engines of the S-1C first stage, which burn kerosene as fuel. The expansion nozzles are 19 ft long and 11 ft 7 in diameter! Each engine was designed to give 1·5 million lb of thrust, equal to that of all eight H-1 engines on the Saturn I first stage, though this has already been up-rated to 1·6 million lb. And each engine burns propellents at the rate of some 3 ton/sec–a total for the vehicle of 15 tons/sec! Remembering that in 1958 the Americans put up a first US satellite weighing 31 lb, the advance in 10 years can be seen to be nothing short of staggering!

Let us now look at the Saturn V vehicle in a little more detail. The first S-1C stage comprises five major cylindrical components. Working from the rear engine end, there is firstly the thrust structure, which weighs 24 tons and is the heaviest of this stage's parts. Then comes the fuel tank containing 203,000 gal of kerosene. Above this is the inter-tank structure which, being unpressurised, has external stringers to help transmit the thrust forces into the next tank section

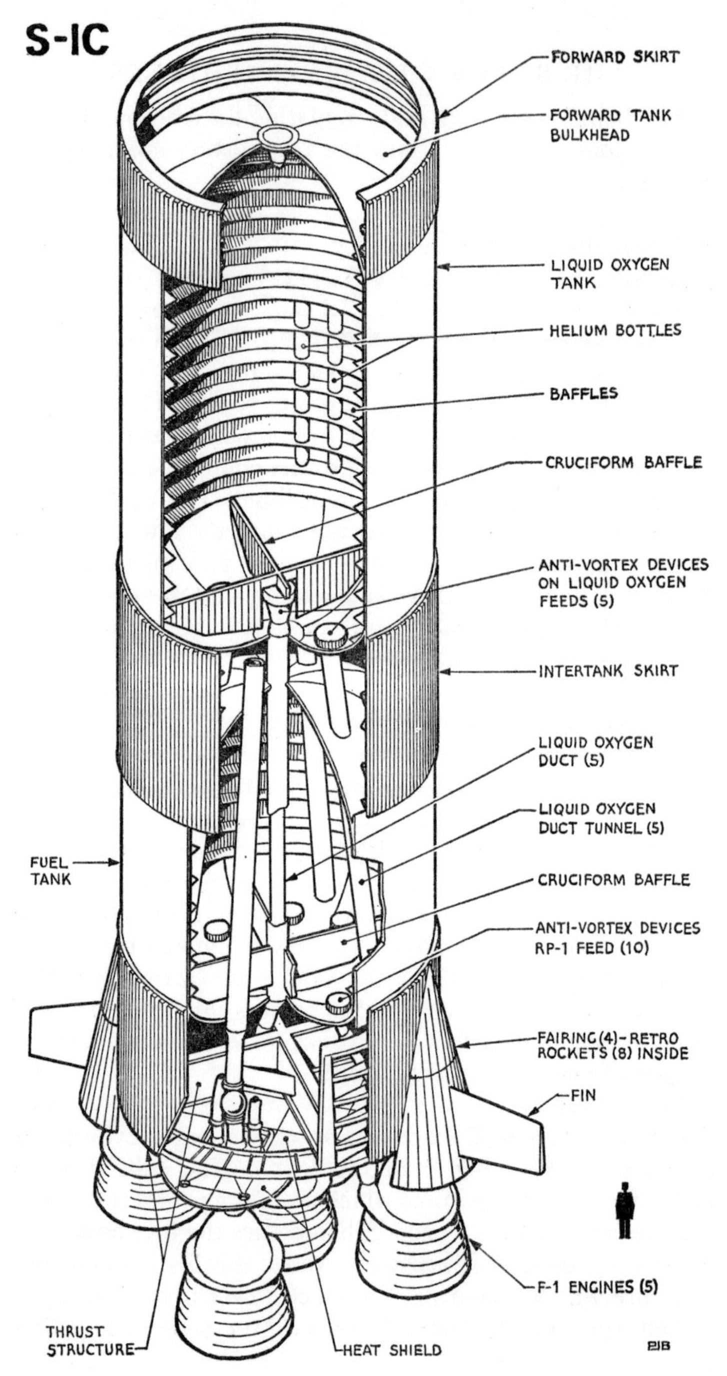

Fig. 23. Saturn V, first stage. Note man for comparison of size.

and to assist in the longitudinal stability and rigidity. The tail section round the propulsion bay is similarly strengthened. Ahead of the inter-tank section is the liquid oxygen tank containing 331,000 gal of liquid oxygen. Finally, ahead of this is the forward skirt, again unpressurised and strengthened with stringers, which is attached to the second stage.

These sections, when seen in Fig. 23, look quite simple and straightforward, but the sheer massive size of the parts created quite extraordinary production problems. The diameter of the tanks, for instance, is 33 ft, and the size of the equipment needed to hold the parts together for assembly and welding can be imagined!

Some of the external geometry should be noted also. The conically shaped fairings at the bottom of the stage extend below the stabilising fins, and are to prevent the full dynamic pressure from the air running along the side of the body from impinging on the parts of the engine nozzles which project farther out than the 33-ft-diameter circle of the stage body. These engines boost the Saturn V vehicle to a speed of some 6000 miles/hour and these fairings ensure that the airflow is deflected and that the shock waves do not produce such a loading as would cause complications with the control system–bearing in mind that each of the outer four engines is gimballed and controlled by hydraulic jacks to provide the steering capability of the vehicle.

The propulsion bay itself is a large and complex structure, which takes the complete $7\frac{1}{2}$ million lb of engine thrust up through into the tank section. To an observer standing underneath a completely assembled vehicle, looking up through some of the access doors before the main fire wall is completed across the base end, there is an impression of looking into the engine room of a battleship, such is the size and space.

The rear kerosene tank is penetrated by double-walled pipes, one to each engine, which carry the liquid oxygen contained in the forward tank through to the inlet side of the main turbo pumps. The pipes are double-walled for insulation purposes, the liquid oxygen being very cold. The walls of the tank are stiffened by internal frames which also act as antislosh baffles. The walls carry a variety of sensors and probes to determine the state and content of the propellent. The liquid oxygen tank is similar, but longer.

The first stage burns for about 150 seconds, though the centre engine is cut off somewhat before this time is up, in order to minimise the forward acceleration, which otherwise continues to increase as the fuel used decreases the mass to be accelerated.

The second stage of the vehicle is known as the S-II stage, and it is built by the Space Division of North American Rockwell Corporation, at Seal Beach, California. The rear part of this stage is again an externally stiffened and unpressurised structure, which contains the five J-2 engines arranged in a similar configuration to those of the first-stage engines–four in a square, with one in the centre. This time, however, the J-2 engines are taking liquid hydrogen with their oxygen instead of kerosene. The liquid oxygen is contained in a nearly spherical tank immediately ahead of the propulsion bay, and ahead of this is the liquid hydrogen tank. Liquid hydrogen has a lower density than kerosene, and this tank is correspondingly larger in comparison with the liquid oxygen tank. The hydrogen tank is free of internal antislosh baffles, and the liquid hydrogen is carried this time back to the propulsion bay through five ducts, which run up the side of the liquid oxygen tank. This time there is no inter-tank section, as the two tanks have a common bulkhead, which is the top of the oxygen tank and bottom of the hydrogen tank. The temperature difference across this fairly thin bulkhead is 126° F. The pressurisation of the tanks is by helium, from bottles carried in the forward unpressurised bay ahead of the hydrogen tank.

All these tanks and bays are 33 ft in diameter, the same as that of the first stage. The length of the second stage is 81½ ft, compared with the 138 ft of the first stage. This S-II stage is the most powerful hydrogen-fuelled launch vehicle at present in production in the Western world. It weighs, fully fuelled, about 1,033,000 lb and its five engines develop just about the same thrust.

An interesting feature is the provision of five solid propellent 'ullage' rockets, each of which develops 2500 lb of

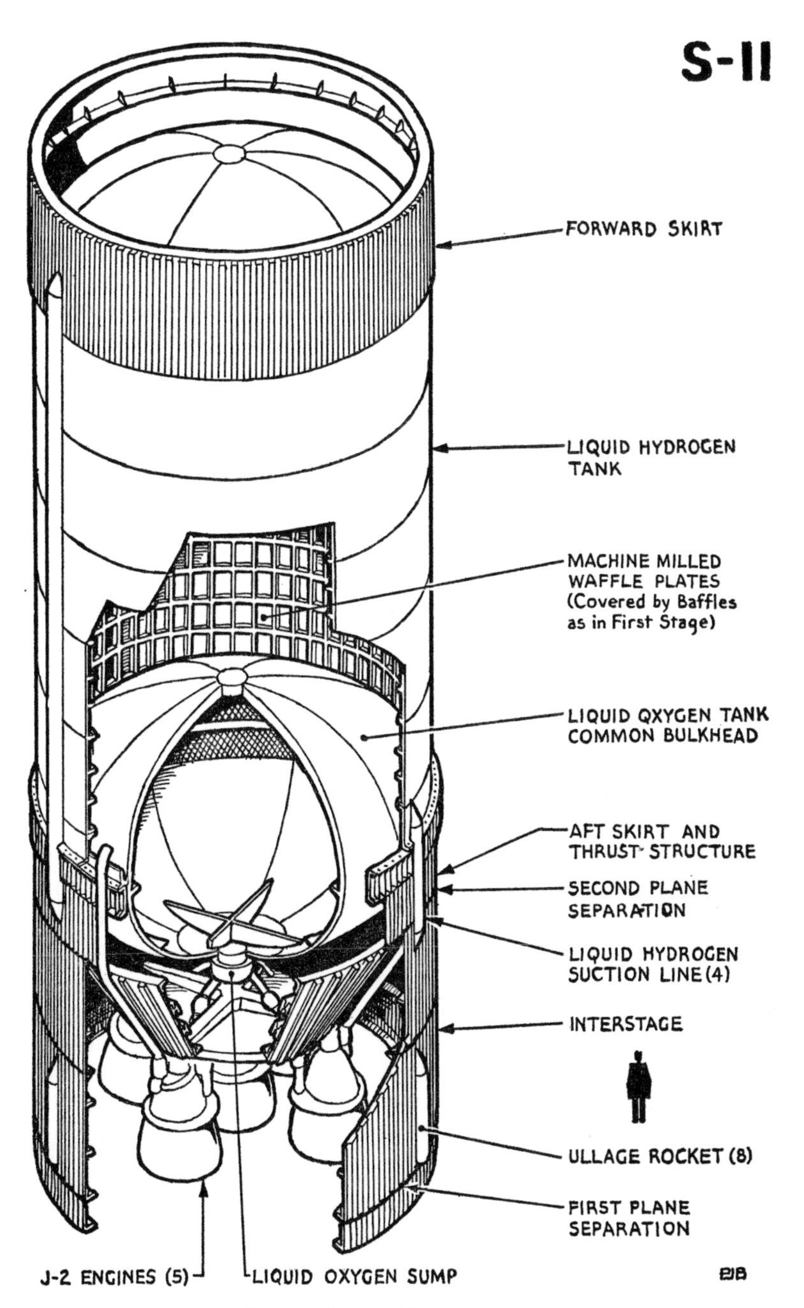

Fig. 24. Saturn V, second stage.

thrust for 4 seconds. The purpose of these is to provide a short pulse to move the propellents to the rear (thus creating an ullage space) which ensures that there is effectively an artificial gravity feed after the first-stage separation so that the fuel system is properly primed ready for the second-stage engine firing.

The third stage of the Saturn V vehicle is the S-IVB, built by the McDonnell-Douglas Corporation at Huntington Beach, California. This is the larger and more powerful successor to the S-IV that served as the second stage for the Saturn I, and is only slightly modified from the S-IVB used on the Saturn IB. It has a tapered aft inter-stage, reducing from 33 ft to 21 ft 8 in diameter, an aft skirt, a thrust structure, propellent tanks and a forward skirt. Liquid oxygen and liquid hydrogen are again used to feed the single J-2 engine, which is identical with the propulsion units of the second stage, and interchangeable with them. This engine is gimballed for main lateral control. Again there are ullage motors, two this time each of 3400-lb thrust, to ensure that the propulsion system is primed after the separation from the S-II stage. There is a general similarity between the construction of this stage and the previous one, the third stage again containing a relatively small and nearly spherical liquid oxygen tank at the rear with the liquid hydrogen tank ahead of that with a common bulkhead. There are some antislosh baffles located in this stage, again in the form of circular frames inside the tank wall. Inside the liquid hydrogen tank are located the helium bottles used to provide the pressurisation for the propellent tanks. Helium is used in all the rocket stages because it is an inert gas–that is, it does not combine chemically with the materials it contacts–and because it remains gaseous at the very low temperatures of the liquid fuels used.

Ahead of the third stage is the instrument unit. This is a small cylindrical section, 21 ft 8 in diameter, but only 3 ft deep. However, its components make up the 'brain' of the Saturn V vehicle, and these have their own environmental control system. Refrigerated 'cold plates' are part of a thermal conditioning system which removes heat by circula-

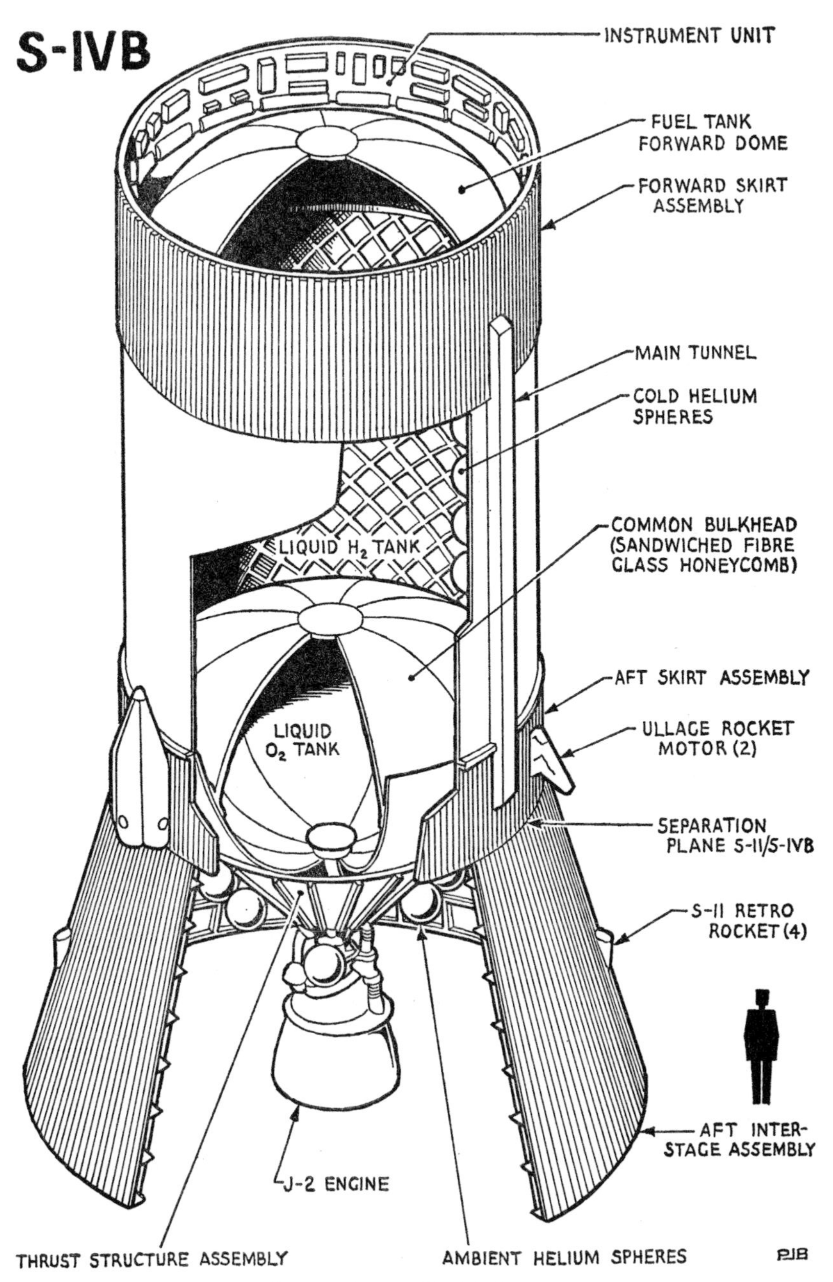

Fig. 25. Saturn V, third stage.

tion of fluid coolant through a heat exchanger, itself evaporating water into the vacuum of space from a separate supply.

The six major systems of this unit are structural, thermal control, guidance and control, measuring and telemetry, radio frequency and electrical. The unit carries out the following functions; navigation, guidance and control of the vehicle; measurement of vehicle performance and environment; data transmission between the vehicle and ground stations, in both directions; radio tracking; check-out and monitoring of vehicle functions; generation and network distribution of electrical power for system operation; and preflight checkout and launch and flight operations. International Business Machines Corporation (IBM) is prime contractor for this very complex unit.

As the Saturn V vehicle rises, signals generated in the instrument unit initiate the separation of the appropriate stages. In a typical case the engines of the finished stage cut-off, and a few seconds later the stages separate by charges being exploded. At the same time the retro-rockets on the finished stage ignite to slow it down, while the ullage rockets on the next stage are ignited to settle the tank contents. The main engines of this next stage are then ignited, building up to full power in 8 seconds or so.

The explosive charges are shaped and run right round the circumference of the vehicle where the separation is to take place and they are, of course, detonated electrically.

It seems rather natural to assume that the stages' separation and assembly–or field splicing–planes are the same. This is rarely so. For instance, on the S-IVB stage, the conical aft skirt on assembly is part of this stage, but on separation it falls away with the S-II stage. The separation of the S-IC and S-II stages is initially very close to the original field-splice plane, but a few seconds later when the ullage rockets have done their work, there is a 'second plane separation' which detaches that part of the S-II skirt carrying these rockets.

There are a number of other explosive devices aboard the stages which come under the heading of 'Propellent Dispersion Systems'. There is one for each stage and they are under the control of the Range Safety Officer. In the event of an emergency arising, such as a vehicle going out of control, coded signals lead to the shutting down of the engines and the explosive ripping open of the tanks to disperse the propellents. Without such safety precautions, especially with vehicles of the size of the Saturn family, one can imagine the holocaust a fully fuelled rocket could cause by crashing.

Another important part of the complete vehicle in its Apollo configuration is the Launch Escape Tower, which in case of emergency can be used to lift the Command Module away from the booster vehicle. This is quite a complex rocket system in its own right. As Fig. 26 shows, it comprises principally a launch-escape motor, which is a large solid propellent rocket, connected to the Command Module by a latticed tower. The whole thing is 33 ft long and weighs about 8000 lb, having a diameter of 4 ft at the largest point. Right at the front is a ballast compartment, the nose cone of which contains instruments to monitor the performance of the whole assembly. Immediately behind that are two 'canard' fins, which are wing-like surfaces deployed 11 seconds after the launch-escape motor is started. These are used for turning the Command Module round so that its base is pointing away from the line of flight. Behind that is a pitch-control motor and this is used in conjunction with the canards to take the whole assembly into a trajectory away from the line of flight of the Saturn V vehicle. The 'whole assembly' consists of the tower and Command Module in the case of a mission termination, or just the jettisoned tower in the case of a successful launch.

When not used, the Launch Escape Tower is jettisoned at 295,000 ft, and this is the maximum height at which the tower can be used. The tower-jettison motor is a solid propellent motor housed behind the pitch-control motor. It is 4½ ft long and burns for 1 second with a thrust of 31,500 lbf. Behind this is a cylindrical casing of 26-in diameter containing the main launch escape motor, which is 15½ ft long and weighs 4700 lb, providing a thrust of 147,000 lb at sea level. Notice that this is twice the thrust of the Redstone rocket which put up the first Mercury capsules–so this

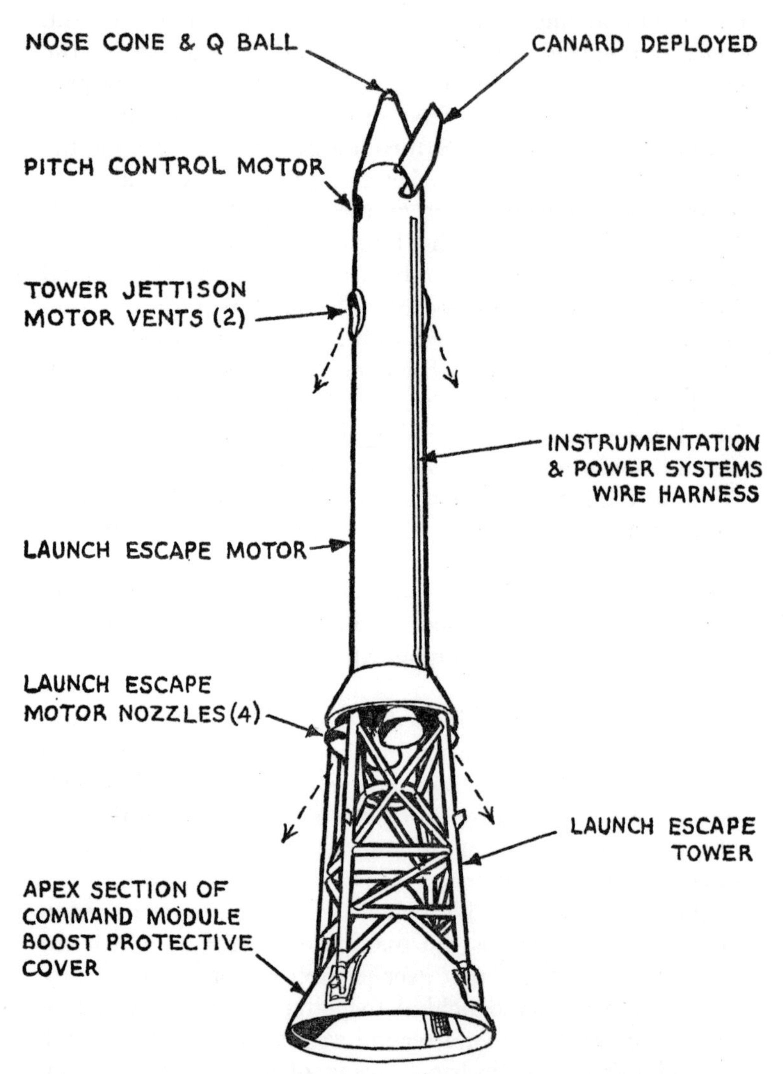

Fig. 26. Apollo Launch Escape System. The Q-Ball measures aerodynamic pressure at the nose.

'small' rocket on the top of the Command Module is 'no chicken'. The exhaust from this motor is passed through four nozzles canted 35 degrees away from the centre-line of the vehicle to take the high temperature efflux away from the front end of the Command Module.

To assist in keeping the spacecraft cool, there is a 10-ft-high welded titanium tubular truss structure, which holds the launch-escape motor away from the Command Module surface. Under this tower is a boost protective cover made of layers of impregnated fibre-glass, honeycomb-cored laminated fibre-glass and cork. It has twelve 'blow-out' ports for the reaction control engines since it completely covers the conical surface of the Command Module to prevent charring of external surfaces during the boost period–remembering that a speed of over 5000 miles/hour is reached while still in the atmosphere–and it is, of course, jettisoned at the appropriate time along with the Launch Escape Tower.

It is of interest to note that to assist carrying the Command Module away from the flight path, in the case of a mission termination, the throat area of one of the two nozzles in the pitch plane is about 5% larger than the two nozzles in the yaw plane, and the second nozzle in the pitch plane is 5% smaller than the other two. The thrust pattern is, therefore, not symmetrical but has a nominal thrust vector angle about 2·75 degrees away from the centre-line through the tower.

Considerable testing has been carried out on this launch-escape system, and it can be used at any time between lift-off –or, indeed, prior to lift-off should some problem occur at the launch pad–and the nominal jettison time of 30 seconds after ignition of the second stage.

Finally, a few facts which might help to put this huge rocket vehicle in perspective. If the Saturn V with its Apollo spacecraft were placed horizontally on the football pitch at Wembley, with the F-1 engines in one goal-mouth the launch-escape tower would poke through to the net at the back of the other goal. Four London underground tube-train tunnels could be run through the propellent tanks with a fair amount of space to spare. The five F-1 engines would devour the contents of a normal road tanker in about 1½ seconds, and in developing their thrust of over 3300 tons they generate the equivalent of 160 million hp–equivalent to sixty-two Concorde supersonic airliners at maximum performance. And while on television most rocket launches look much alike, when the Saturn V takes off a mass of 2650 tons–equivalent to a large modern naval destroyer–is leaving the Earth on the most fantastic piece of controlled power man has yet devised.

TABLE I

Sequence of Events at Saturn V Lift-off (Typical)

Time from lift-off in hours: minutes: seconds	
00:00:00	Lift-off
00:01:17	Maximum dynamic air pressure: beyond this point, although speed increases, air density falls off rapidly
00:02:06	S-IC stage centre engine cut-off: this is to prevent the acceleration from increasing beyond a value set as suitable for the astronauts
00:02:31	S-IC stage outboard engines cut-off. Height, 42 miles; distance down range, 54½ miles; speed, 6070 miles/hour
00:02:32	S-IC/S-II stages separation: S-IC retro rockets and S-II ullage rockets ignition
00:02:33	S-II stage engines ignition
00:03:07	Launch Escape Tower jettison
00:08:40	S-II stage engines cut-off. Height, 122 miles; distance down range, 927½ miles; speed, 15,258 miles/hour
00:08:41	S-II/S-IVB stages separation: S-II retro rockets and S-IVB ullage rockets ignition
00:08:44	S-IVB stage engine ignition
00:09:00	S-IC stage splashdown, 391 miles down range
00:11:32	Insertion of S-IVB stage and spacecraft into parking orbit. Height, 120 miles; speed, 17,500 miles/hour
00:20:00	S-II stage splashdown, 2600 miles down range
02:50:30	S-IVB stage engine ignition for injection into trans-lunar trajectory. This time varies according to the point in the parking orbit where the ignition is made.
02:55:43	S-IVB stage engine cut-off
03:09:14	S-IVB/Command and Service Modules separation. Procedure from here on varies with the mission.

TABLE 2

Apollo Spacecraft and Rocket Vehicle Weights (Typical) Pounds

	Dry weight	*Consum-ables*			
Launch Escape System					8,200
Command Module	11,700	1,300		13,000	
Service Module	11,500	43,500		55,000	
Lunar Module					
Ascent stage	4,200	5,900	10,100		
Descent stage	3,800	18,500	22,300		
			———	32,400	
				———	100,400
LM Adapter					3,900
Saturn V					
Instrument Unit				4,750	
3rd stage	26,000	238,250		264,250	
Interstage				8,750	
2nd stage	88,600	946,900		1,035,500	
1st stage	305,700	4,483,300		4,789,000	
				———	6,102,250
					6,214,750

10

Launch Sites and the Mobile Concept

EARLY on in the programme it had become necessary to decide upon a suitable site for the Saturn rocket launchings. Seven possible sites scattered over the United States were investigated, though there were obviously special advantages in integrating this new launch facility with an existing rocket range. Apart from Wallops Island in Virginia, where parts of the Mercury system had been tested, and some sites for sounding rockets, there were three ranges in the USA.

First, there was the US Army's White Sands Missile Range utilising some of the waste desert land in the interior. However, with the increasing size of launch vehicles, this range was already reaching its limit. On the Pacific coast, about 50 miles north of Los Angeles, was the US Navy's Pacific Missile Range with facilities for firing the big rockets of the time, such as the Atlas and Thor. As firings were out over the Pacific, there were no restrictions on the size of vehicle which could be launched. Thirdly, over to the east in Florida was the Atlantic Missile Range operated by the US Air Force. Geography rather favoured this site as a rocket range. Firings could be made in a south-easterly direction over 5000 miles of the Atlantic Ocean. While there was little chance, therefore, of a crashing booster doing any harm, this line of fire passed close to the West Indies string of islands and to Ascension Island in mid-Atlantic. Twelve tracking stations had already been constructed on these islands so that rocket vehicle launches could be accurately monitored over all the critical launch stages.

There was one other factor which favoured Florida as a site–the direction of firing. The Earth spins on its axis from west to east, and a rocket sent up on the equator already possesses a west to east motion of 1500 ft/sec or just over

1000 miles/hour. A rocket, therefore, sent off in an easterly direction starts with a bonus of about 6% of orbital velocity; this is a 12% gain over one sent up in the opposite, that is westerly, direction. At the latitude of the main American missile ranges this bonus is reduced to about 1300 ft/sec or $5\frac{1}{2}$% of orbital velocity, and on a south-easterly course as

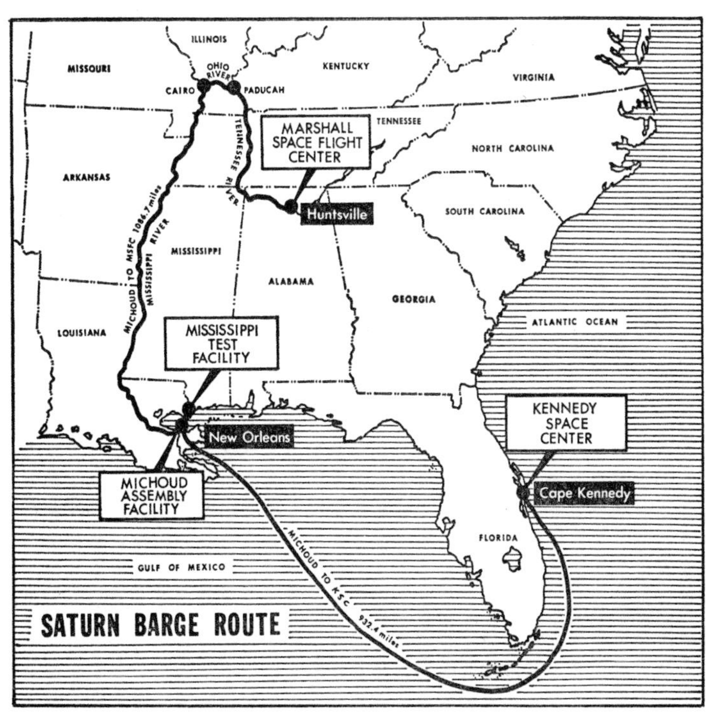

Fig. 27. This map of the Saturn vehicle barge route also shows the geographical locations of the major Apollo facilities sites.

would be the case from Florida this is again reduced to under 4% or about 650 miles/hour. Nevertheless, this is a sizeable bonus when dealing with a very large rocket, when working towards the limits of current technology and resources, and when there are few margins to play with. By comparison, firings on the other ranges were either north, south or west.

Although, perhaps, not ideally situated, the Florida site also fitted in very well with the pattern of transportation being worked out. Rocket vehicles not only have to be built but they have to be transported from the manufacturers' factories to the launch site, since it is impracticable to consider resiting the factories with their special equipment and the personnel with their special skills. At quite an early stage it was apparent that the rocket-vehicle sections to be transported would be between 20 and 35 ft in diameter and up to 150 ft long, and a feasibility study showed that movement by water was the only practicable answer to the transport problem. The Cape Canaveral area of Florida fitted in well with this concept, being able to accept both sea-going ships and river barges.

The latter may sound strange, because on a small-scale map the site appears to be just a port on the coast. In reality the east coast of Florida is made up of a chain of small islands, the spaces between them and the mainland being called and treated as rivers.

Indeed, the actual site chosen to be America's 'Moonport', and later officially called the John F. Kennedy Space Center (abbreviated to KSC), was on Merritt Island, about 10 miles north of Port Canaveral. The Indian River, or Intracoastal Waterway, separates it from the mainland; to the east is an ill-defined area of water called the Banana River and a strip of land varying in width from a few miles to a few hundred yards. Beyond this is the Atlantic. NASA acquired 88,000 acres here for building the KSC facility on a site which was near wilderness. Much of the area was sodden, sandy marshland, the remainder being covered almost exclusively with sparse pine, palm trees and palmetto scrub, together with a variety of wild shoulder-high vegetation associated with the less fertile sub-tropical coastline area of this region.

The part of this Center associated with the Saturn V vehicle is called Complex 39 and, as the form this has taken has broken precedent, being considerably different from previous layouts and practice, it will be as well to look briefly first at two other launch sites a few miles south, known as Complexes 34 and 37 from which Saturn I vehicles have been fired, and which were a direct extension of current practice.

When the early V2 rockets were being readied for firing,

they stood upright on their concrete launch pads, and the technicians conducted their last-minute checks and servicing standing on fire-escape ladders. As rocket vehicles grew in size, these ladders were replaced by wooden platforms, and then with removable gantries or service structures, which continued to become ever bigger and more complex. Also, until the rocket engines are working and can supply auxiliary

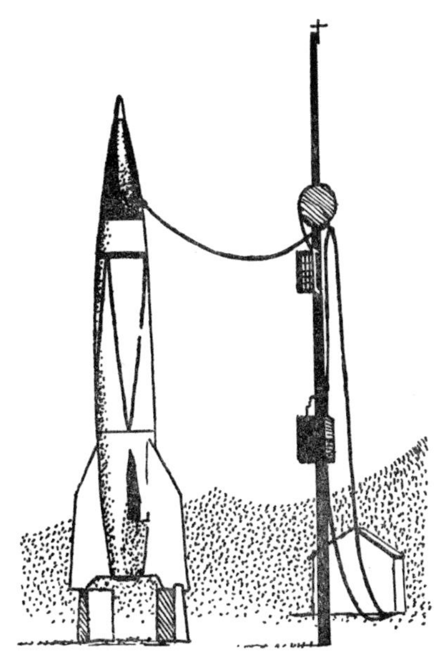

Fig. 28. V2 with Umbilical Mast at White Sands, New Mexico.

power for on-board equipment, electric, hydraulic and pneumatic power has to be supplied from external ground sources. The cables used for this purpose were, in early models, slung from a mast and detached themselves from the rocket as it rose. Because of the analogy with a child being born, these attachments became known as umbilicals. As rockets continued to grow, it became necessary to provide all kinds of services to all the many rocket stages and, as well as putting in fuel, power, air-conditioning, information and command signals, a lot of information had to be taken out of the launch vehicle about the state of many parts and systems for the benefit of the personnel controlling the launch. The number of temporary attachments grew until the umbilical mast had developed into towers of considerable size and complexity.

We are now in a position to have a look at the general form of Complex 34, which was completed in time for the first launch of a Saturn I vehicle on 27th October 1961. The first four launches in the Saturn development programme took

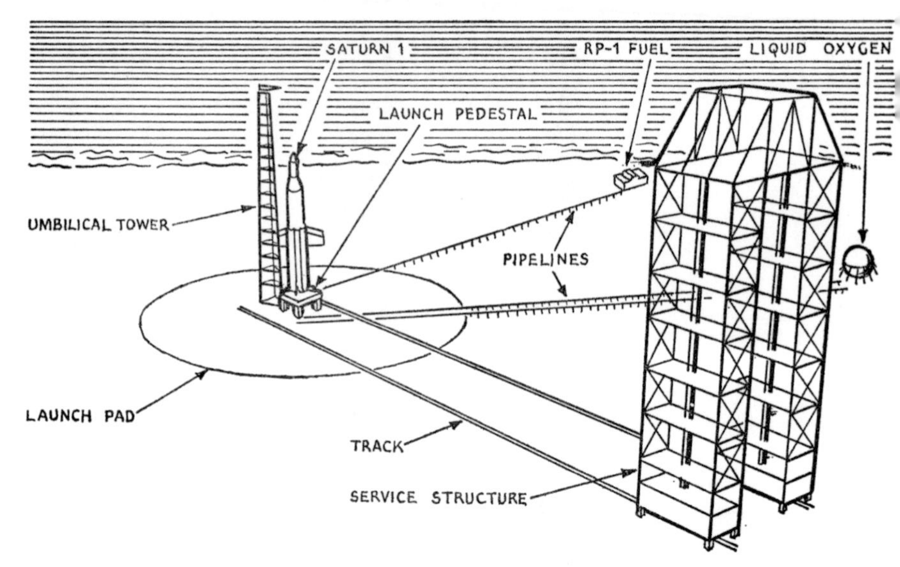

Fig. 29. General arrangement of Complex 34 at Kennedy Space Center. The Blockhouse is off right.

place from this site, the last one being on 28th March 1963. While the remaining Saturn I vehicles were being launched from the twin site, Complex 37 just to the north, Complex 34 was modified for launching the enlarged Saturn IB. Fig. 29 shows the general layout of this site.

In the centre is the launch pad, a concrete base 430 ft in diameter, part of which is covered with refractory brick to minimise damage from the rocket exhaust.

On this stands the 42-ft square reinforced concrete launch pedestal. Positioned in the centre of the pad, this provides a platform for the launch vehicle and certain ground-support

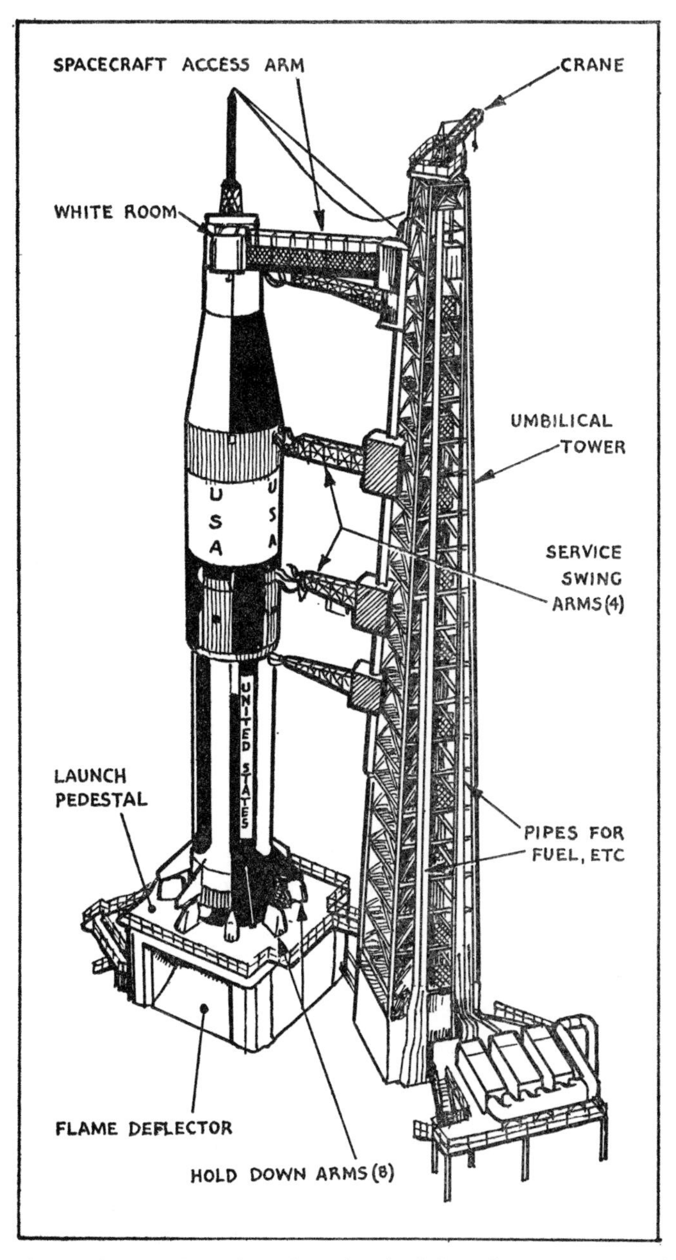

Fig. 30. Saturn IB on launch pedestal of Complex 34 at Kennedy Space Center.

equipment. Steel-plate covers all the surfaces exposed to the rocket flames, which are exhausted through a 25-ft diameter opening to a deflector below. Here the exhaust flames are split in two and turned through 90 degrees to emerge horizontally left and right.

Next to this is the 250-ft-high umbilical tower, a steel-trussed structure with four swing arms. These arms are each quite heavy structures in themselves. They hinge about pivots on the tower and, in the extended position, each arm carries links between the space vehicle and the tower leading to ground-based power, air-conditioning, hydraulic, pneumatic, fuel, measuring and command systems. As the rocket begins to rise off the pad these arms swing away to leave the path clear, the broken connectors at the ends of hoses, etc., being made self-sealing.

As this facility is used for launching the Apollo spacecraft into Earth orbit as part of the development programme, at the 220-ft level on the tower an Apollo spacecraft access arm is fitted. The astronauts go to and from the spacecraft through this access arm at the end of which, abutting against the spacecraft, is the 'white room', a very clean, air-conditioned area. This arm is reached by the umbilical tower lift which can move at 450 ft/min and is the astronauts' primary escape route in emergency. A slide-wire system provides an alternate means of quick escape from the pad area. It takes only 30 seconds for a man to travel the 1200 ft down the slide-wire from the 220-ft tower level to the edge of the launch complex.

Next comes the service structure, a movable steel framework used during vehicle erection, assembly and checkout. It provides work platforms for personnel, cranes for lifting the various rocket stages and the spacecraft into place on the launch pedestal, and protection from the weather for both the space vehicle and the launch personnel. During its modifications to suit the bigger Saturn IB vehicle, hurricane gates were installed on this structure so that the rocket could ride out hurricane-force winds without having to be moved to a hangar area–a time-consuming operation introducing further risks.

Within the structure legs there are four lifts and seven fixed platforms at various levels, and eight enclosed platforms can be extended to the vehicle from the tower. Two further work platforms located near the top of the service structure allow access to the launch-escape system fitted on top of the Apollo spacecraft.

The whole service structure weighs 3552 tons and moves on four 12-wheel trucks along a special dual-track railway within the complex until it reaches the launch pad, when support points remove the service structure from the trucks and effectively anchor it to the ground. The structure is moved back 600 ft to its parking position before the rocket is launched, a 500-kVA diesel-electric generator, enclosed in the base, powering the 100-hp traction motors in each truck.

About 1000 ft from the launch pad is the blockhouse, which houses the launch team, instrumentation and control equipment. Because of its proximity to the launch pad and the rocket's $1\frac{1}{2}$ million lbf lift-off thrust, this blockhouse has been designed to withstand blast pressure of up to 2200 lbf/in^2 and the walls vary from 7 ft thick at the top to 30 ft thick at the base. The building is dome shaped and contains 11,650 ft^2 of space on two floors. The top floor houses launch control and the various monitoring and recording consoles, while the ground flour contains one of the computers used for automatic checkout and the personnel involved in tracking, telemetry, closed-circuit television and communications. During the final count-down for the Apollo 7 mission in which the Command and Service Modules were tested in Earth orbit, the blockhouse was manned by a crew of about 250.

In order to understand better the design philosophy employed in the Saturn V Complex 39, it is useful here to study some of the work and operations carried out on this Complex 34 pad for the Apollo 7 mission.

The two propulsion stages–the S-IB and S-IVB–of the Saturn IB vehicle were erected on the pad in April 1968, after which a number of preliminary tests were carried out on the individual stages prior to electrical matching up and conducting integrated systems tests of the overall vehicle.

The Apollo 7 Command and Service Modules arrived at

KSC in May 1968 and preliminary checking of spacecraft systems was carried out at the Manned Space Flight Operations Building. The two modules were put together in this building's vacuum chamber for a series of unmanned and manned altitude runs. The spacecraft was then fitted to its adapter and transported to Complex 34, where it was mated to the launch vehicle in August, mechanically at first with the electrical matching up taking place 3 weeks later, after which overall tests of the integrated launcher and spacecraft could begin.

The launch-escape tower on the top of the Command Module was fitted and a series of simulated missions were performed. One of the major tests among this series was made with umbilicals disconnected in a complete launch mission rehearsal called a 'Plugs Out Test'.

A Countdown Demonstration Test was also conducted about a month before the 11th October scheduled launch date. This was a complete dress rehearsal of the count-down, including a 'wet' test when the Saturn IB vehicle was fuelled up, although the astronauts were not in the spacecraft during this part of the test.

About 2 weeks before launch, a Flight Readiness Test was conducted to exercise the launch vehicle and spacecraft systems, this time Mission Control Center at Houston participating in the test together with the local team.

The original designs for Complexes 34 and 37 were put in hand about 1959, soon after NASA had been established and the form of Saturn I was becoming apparent. When it came to planning the Complex 39 facilities, a situation had arrived when it was necessary to do some new thinking and reexamine the methods until then adopted.

Up to this time, the stages of rockets had been assembled on their launch pads. However, even with the Saturn IB vehicles a pad could be tied up for 6 months or so while all the necessary assembling, testing and count down checks were carried through. The increasing complexity and sophistication of the equipment fitted into the rocket stages had already led to the enclosing of working platforms and sometimes even to local air-conditioning. More and more equipment was also needed on the spot for the many assembly and checking operations. Indeed, the service structure, especially with hurricane doors fitted, was becoming virtually a moving building and in future would probably have to be thought of in this way. Undoubtedly the situation would be even more severe in the case of the Saturn V vehicle. It had three instead of two stages and would need effectively factory conditions for much of its assembly and testing–and going by experience to date, 6, 9 or even more months might not be unreasonable a time between reception of rocket stages and an actual firing. If a series of planned development launches were to be made at, say, 3-monthly intervals, perhaps four very large launch pads would be required with a duplication of many complex facilities at each pad.

Consequently, in order to cut down on the land area required, and to gain more efficient usage of manpower, equipment and buildings, it was suggested that it might be more expedient to build up the Saturn V rocket vehicles in a centralised hangar or building and only move them out to the launch pad a few weeks before the actual firing. This idea was followed up and became known as 'the mobile concept'.

Once it was determined that the advantages of the mobile concept far outweighed those of previous launch modes, the means of implementing the idea had to be devised. This called for the preliminary design of the launch pads, the assembly building and some means of moving the launch vehicles from the assembly building to the pads.

To begin with, it soon became apparent that it would be impracticable to move the large Saturn V rocket on its own. During assembly and testing on previous rocket vehicles, so many connections were made between them and their umbilical towers–connections which were continued in many cases right up to the moment of firing–that it seemed necessary to create mobile launch towers and bases upon which to assemble and fire the rockets. It would also be necessary to provide a more simple kind of service structure which could be moved into position during the Saturn V's few weeks on the firing pad.

The Saturn V empty weight with the Apollo spacecraft

would be somewhere between 400,000 and 500,000 lb–about 200–250 US tons; however, fuelled up this weight would increase to 6 million lb or 3000 US tons. While the full weight would not have to be transported, the mobile launcher would have to be strong enough to support this load and take the extra forces occurring during blast-off. Going by Saturn IB precedents, it seemed likely that this launcher could weigh up to 6000 tons. The transporter for this load could also turn out to be in the region of thousands of tons weight. The designers were thus faced with moving over land a mass of around 10,000 tons over a distance of, perhaps, 3 miles.

One of the key elements in this mobile concept was indeed the feasibility of carrying out such a transport operation. Several means of transport were considered, some of them quite imaginative. Among the methods originally examined in some depth were a pneumatic-tyre transporter, a hovercraft or ground effect machine, a rail system and a barge system. In an effort to capitalise upon and exploit the best features of each, a hybrid system was even considered. Most of these modes were prone, however, to severe drawbacks and many would require extensive and time-consuming development beyond the current state-of-the-art. The time schedule of the whole Project Apollo operation was such as to rule out untried ventures.

The complexity involved in the selection and development of special multiple steering mechanisms, scuff-free rolling turns, and load levelling control problems discounted the pneumatic-type transporter. The hovercraft principle transporter was rejected because a critical problem at that time, especially for achieving satisfactory power requirements for the load, was that of developing suitable seals against peripheral air leakage–skirts as they are called today. Such a craft, without special constraints, would have no positive contact with the ground and would be sensitive to wind loads and inclines. Indeed, the main cause for rejection lay in the complexity of the 'roadways' that would be needed to use the hovercraft principle in this particular application. The rail system failed because of the engineering problems that arose from a vehicle with an enormously wide wheelbase. A large number of wheels would be required to spread the load and the necessary roadbed again would become most complex. The barge system was shelved because of inherent roll instability with such a tall load, which would adversely affect both the launch vehicle and the control of the system. However, in this case propulsion proved to be one of the main difficulties, coupled with the enormous cost involved in developing a suitable launch pad to match up with water-borne barges.

Rather providentially, it came to notice that the world's largest mobile land machine–a huge 8500-ton stripping shovel–was to become operational during 1962, this gigantic machine being developed for surface coal-mining in Kentucky. It was immediately recognised that this machine represented the existing state-of-the-art in this area, and if the principles could successfully be adapted reasonably low development costs would accrue.

The machine in question was the Bucyrus-Erie Company's 3850-B Shovel being built for the Peabody Coal Company. The 'dipper' which actually did the shovelling had a capacity of 115 yd^3–the size of a small house–and the machine boom and arm had a reach of 420 ft, shifting 200 tons of material at each dipper bite. The machine, except for the dipper arm, was structurally rather like a very large turntable crane standing on four supports, each of which was carried on a twin-crawler truck. The total weight carried by the near three-hundred pads of the crawler tracks was between 8500 and 9000 tons–very similar to the load expected to be carried at Complex 39.

This massive machine used proven crawler principles for mobility and proven jacking principles for levelling purposes, and it became apparent that load-bearing capacities of considerable magnitude could be designed into such a system. As a result, the Crawler-transporter idea–as it became known–was accepted on 25th July 1962 and the engineering teams could then proceed with the assurance that the mobile concept for Complex 39 was a feasible proposition.

Briefly, then, the operations could be envisaged as follows.

Rocket sections and other heavy or bulky parts would be brought in by water up the Banana River and would be unloaded on to special transporters for moving into the Vehicle Assembly Building. Here these parts would be built up on a Mobile Launcher platform and all systems would be tested to an advanced state. The Crawler-transporter would then move the Mobile Launcher with its Saturn V rocket vehicle

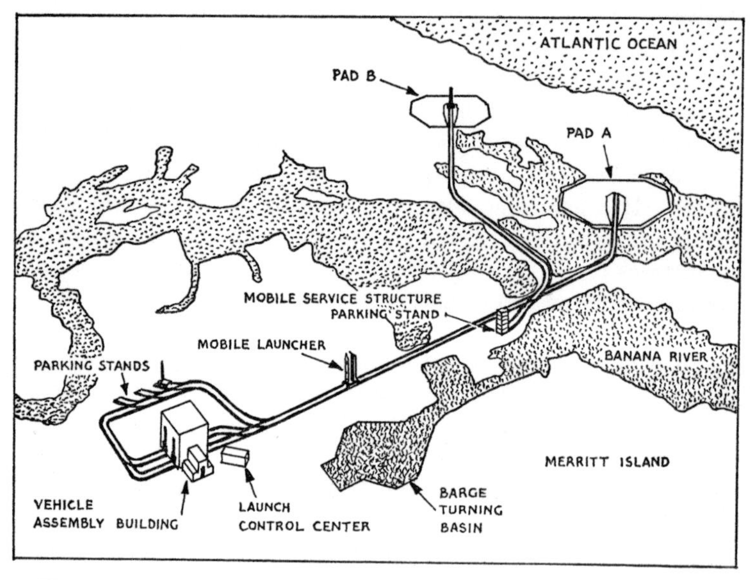

Fig. 31. General view of pads A and B of Complex 39 at Kennedy Space Center.

out to the Launch Pad, some 3 miles away, after which it would return to collect a Mobile Service Structure so that the rocket vehicle had adequate servicing facilities for its last phase on the pad. Just before the Saturn vehicle was due to be launched, the Crawler-transporter would take the Mobile Service Structure back to its parking stand, and after the launch it would remove the Mobile Launcher for refurbishing ready for another cycle.

The layout eventually worked out for this Complex is shown in Fig. 31. This provided for two Launch Pads about

3 miles from the Vehicle Assembly Building and, of course, took into account the local topography.

Before we examine the development of the launching arrangements in more detail, it will be useful to get into focus the overall organisational set-up now that the scale of the whole operation is becoming clear.

In the early days of space flight, the management organisation for Project Mercury was the NASA Space Task Group based at Langley Field in Virginia. In October 1961 it was decided to move this group to Houston in Texas to establish the Manned Spacecraft Center for Projects Gemini and Apollo. Already in 1960 the Army Ballistic Missile Agency Establishment at Huntsville had been taken over by NASA to become the Marshall Space Flight Center under Dr von Braun. In the meantime, a large number of research establishments were taking on work connected with Project Apollo, a parcel of land in Florida had been acquired for the Kennedy Space Center, test facilities were being planned and main contractors were taking over large parts of the development work. Of course, the organisational side had to grow in step.

At the top, now, is the President of the United States, who has a number of important committees dealing with science, defence and so on. One of these is the National Aeronautics and Space Council, which administers the programme through the National Aeronautics and Space Administration (NASA) in Washington. For most of its time NASA was headed by James E. Webb, until on his retirement he was succeeded by Dr Thomas Paine. The US Congress has a Senate Committee on Aeronautics and Space Sciences, and a House of Representatives' Committee on Science and Astronautics, but these are not of an executive character and have no direct say in the running of the project.

The NASA Administration has a large number of small departments which need not concern us, such as Policy Analysis, Legislative Affairs and so on. However, under the Administrator's control are also five large departments–Organisation and Management, Manned Space Flight,

Space Sciences and Applications, Tracking and Data Acquisition, and Advanced Research and Technology.

Dr G. E. Mueller has been in charge of Manned Space Flight since 1963 and under him come departments on Program Control, Space Medicine, Management Operations, Mission Operations, Advanced Manned Mission Program, Apollo Applications and, of course, Apollo Lunar Landing Program, the latter being under General Samuel C. Phillips. Also under Mueller are three major establishments–the Marshall Space Flight Center at Huntsville, directed by von Braun, which is concerned mainly with launch vehicle research and development, the Kennedy Space Center under Dr K. H. Debus, dealing with space vehicle launching, and the Manned Spacecraft Center at Houston, under Dr Robert R. Gilruth, which is naturally concerned with manned space flight; it has field offices also at the Kennedy Space Center.

The Marshall Space Flight Center organisation is complex, but it breaks down essentially into two parts, one concerned with Research and Development Operations, which deals with all essential launch vehicle design work, and one concerned with Industrial Operations, which has departments dealing with the day-to-day running of programmes with industrial contractors, including the Engine Development Program, and has also under its wing the Michoud Assembly Facility and the Mississippi Test Facility.

The Michoud plant was designed originally as a shipyard during the Second World War, but it had seen little further use until taken over as a centre for the assembly of the Saturn I and Saturn V first stages. The total complex covers 825 acres and is 15 miles east of New Orleans.

The booster stages are taken by water from the Michoud plant to the Mississippi Test Facility site, some 35 miles north-east, where the rocket stages are mounted in giant test stands for the static firing of the rocket engines. Four test stands were built initially, these being huge concrete structures in a roughly 5-mile-square area embracing 13,500 acres. The site has fee and easement rights over 142,000 acres and the whole establishment was originally expected to cost

around $250 million. This facility is very important, for once a space vehicle is launched it cannot turn back for repairs. The economics of rocketry require that the expensive vehicle's mission should be assured of a very high degree of success before countdown, and these facilities allow complete testing on the ground without losing the rocket stage.

The Manned Spacecraft Center has been under the direction of Dr Gilruth since its creation in 1961 and it carries out essential work in the design of manned spacecraft and deals with astronaut training. Once a manned spacecraft has been successfully launched control is transferred from Kennedy Space Center to this Center in Houston.

Under Space Science and Applications come three establishments, the Jet Propulsion Laboratory, Pasadena, California, which deals with instrumented exploration of the Moon and planets, Wallops Station, Virginia, concerned with sounding rockets and small satellites, and the Goddard Space Flight Center, Maryland, which is generally concerned with unmanned spacecraft and tracking, but whose computers and switching facilities are used during manned flight missions to save duplication.

U der Advanced Research and Technology come the Ames Research Center, California, concerned with life-support systems, the Flight Research Center at Edwards Air Force Base, California, dealing with supersonic flight research mainly with rocket planes, the Langley Research Center, Hampton, Virginia, on materials and structure research, the Electronics Research Center at Cambridge, Massachusetts, and the Lewis Research Center at Cleveland, Ohio.

On top of this basic organisational structure and list of establishments, one must add also the main contractors for the launch vehicles, engines, spacecraft, communications equipment, and ground facilities, and their 20,000-odd subcontractors.

It really is quite remarkable that so many people working under such a large organisation have been able to bring together so much complex, sophisticated equipment at the right time so successfully.

11
The Launch Pad and the Mobile Launcher

In Complex 34 for the Saturn I and IB vehicles, the launch pedestal, upon which the rocket stood, and the umbilical tower were fixed directly to the concrete launch pad, which in that case was essentially a flat circular area 430 ft in diameter. The pads for Complex 39 would need to be considerably more complex, since the pedestal and umbilical tower were to be mobile. Work was started in 1963 with a view to carrying out the preliminary operational work in 1966.

The general form of this new kind of launch pad is shown in Fig. 32. The concrete base, now called the hardstand, is about 40 ft above the ground surface and has sloping embankments, while through the centre lies a flame trench for the rocket engine exhaust flames. The mobile launch pedestal, or Mobile Launcher base as it is now called, stands on supports over this trench, and the umbilical tower is built upon this base.

In the centre of the Mobile Launcher base is a square blast opening over which the Saturn V rocket stands. Below this is the flame deflector, which splits the rocket exhaust flames to pass out at both ends of the flame trench. This flame deflector runs on rails and can be moved out for repair or replacement. Since this flame deflector will not be mentioned again, it is useful to consider it in more detail at this point. It weighs about 650 tons and, although made in metal, the deflection surfaces are covered with $4\frac{1}{2}$-in thickness of a refractory concrete consisting of volcanic ash aggregate in a calcium aluminate binder. The heat and blast of the rocket engines were expected to wear about $\frac{3}{4}$ in from this surface during a Saturn V launch. After the deflector has been moved into position on its rails, the wedge is raised a few feet into position hydraulically.

On each side of the flame trench on the hardstand runs the pair of crawler ways, over which the Crawler-transporter moves to bring the Mobile Launcher to its support pedestals. These crawler ways rise to the hardstand level up a 5 % ramp.

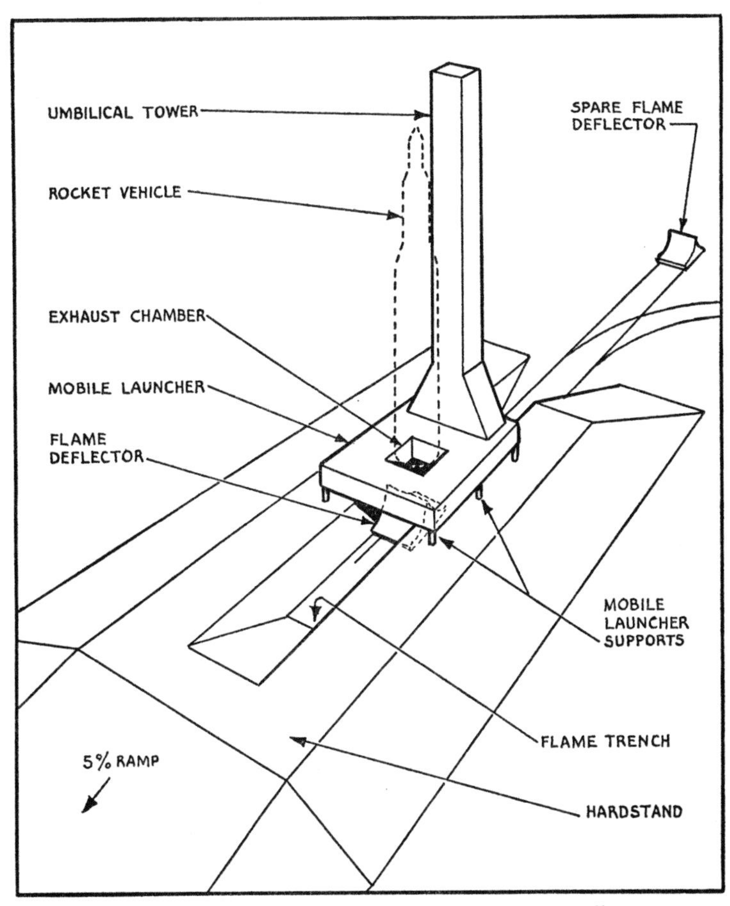

Fig. 32. Preliminary idea for launch pad to suit the mobile concept.

Two Launch Pads, designated A and B, have been built as part of the Complex 39 facilities, the crawler ways from each meeting at a junction and then continuing substantially straight to the Vehicle Assembly Building. Since with this development there is no pad proper in the sense of a circular

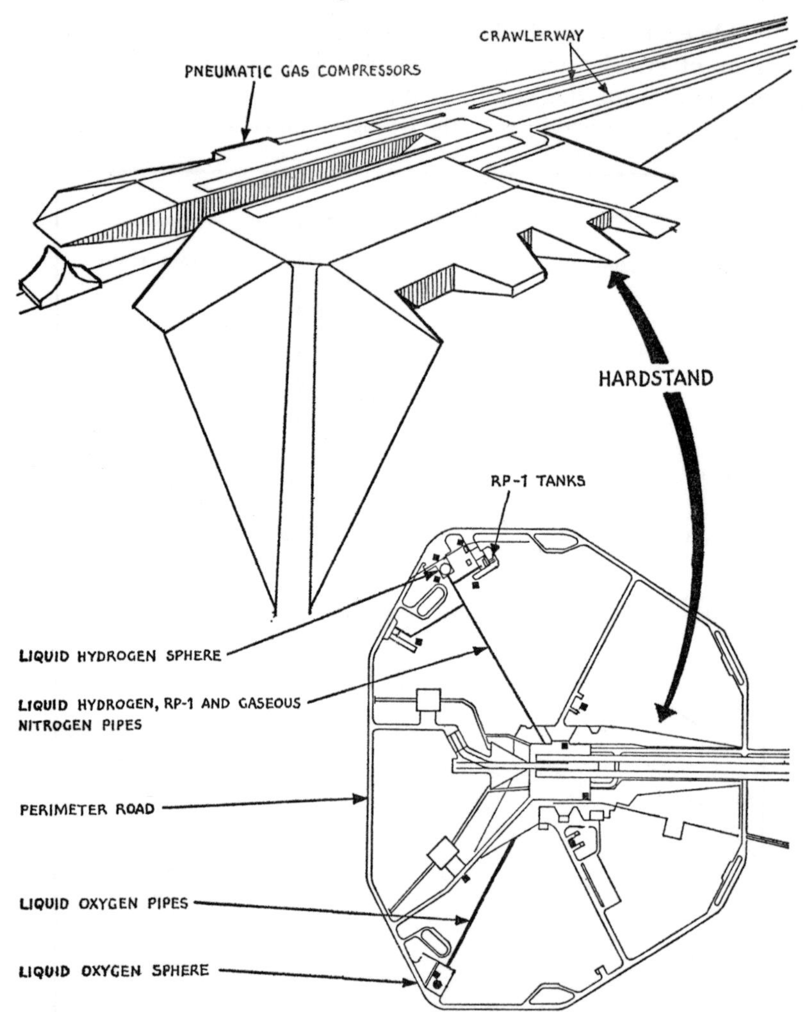

Fig. 33. A Complex 39 Launch Pad.

concrete base as in earlier cases, the term Launch Pad has here been extended to cover the complete local complex. Each Launch Pad in this wider sense is an eight-sided polygon about 3000 ft across, covering about a quarter of a square mile. As seen in Fig. 33 the hardstand is essentially in the centre, while dispersed round the perimeter are the storage facilities for the Saturn V propellents–RP-1 high-grade kerosene, liquid hydrogen and liquid oxygen.

The building of these pads was quite a sight. Each launch pad began as a huge pyramid of surcharge dredged up from the surrounding swampland. Built in three steps or levels, this pyramid was initially 80 ft high. Over a period of 4 months, the weight of this surcharge caused the sub-soil to settle as much as 4 ft. It was, of course, necessary that this settlement should take place at this stage rather than later, especially when one considers the weight to be carried by the sub-soil. Apart from the weight of the hardstand itself and the fixed facilities, these would be supporting at various times the weight of the Saturn V vehicle fully fuelled, say 3000 tons; the weight of the Mobile Launcher, 6000 tons; the weight of the Mobile Service Structure, nearly 5000 tons; and the Crawler-transporter at 3000 tons.

The next stage of development involved the levelling of the top to the correct height for the hardstand, which is about 48 ft above sea level, and the forming of the approach ramp. The task of digging the central flame trench came next and with the pouring of the reinforced concrete eventually the pad began to take on a recognisable form. While it might appear more simple to have designed the hardstand to be at ground level, so that no ramp was required, as the water-table in these swampy conditions was at most only a few feet down, the flame trench would then be flooded to about 40 ft depth.

On the east seaboard side of the hardstand is housed the pneumatic gas compressor facilities, while on the other side of the hardstand is the environmental control facility. It is not possible here to discuss all the equipment and facilities appertaining to these Launch Pads, since they become quite complex though each system is simple in principle. However, there are some interesting features worth noting.

Apart from having umbilical connections between the rocket-vehicle stages and the umbilical tower, now that the tower is part of a *Mobile* Launcher, when this arrives at the Launch Pad facilities have to be at hand for connecting up all the various services from the Mobile Launcher to what have been called umbilical interface towers. These are small support structures on the hardstand which carry the ground ends of the various pipes, hoses and cables. Included here are such things as the liquid hydrogen tower, the high pressure pneumatic tower, the fuel-system tower, and east pad elevator and stair tower for personnel and equipment on the east or seaward side; and on the other side are the liquid oxygen tower, electrical entries, environmental control system tower, west pad elevator and stair tower, facilities and utilities pedestal and electrical power pedestal.

When one considers all the electrical power requirements of the Mobile Launcher, Mobile Service Structure and the Saturn V rocket, all the thousands of instruments monitored in the vehicle, lines for command signals and so on, one can imagine how complicated these interfaces would be to describe in any detail. On the more mundane side, drinking water, washing water and even sewage connections have to be made to the Mobile Launcher and Service Structure; and, as we shall see later, a fair amount of water has to be pumped in for cooling purposes during the actual minutes of rocket vehicle lift-off.

Of special interest are the supplies of liquid oxygen and liquid hydrogen, because oxygen has to be cooled to minus 297° F before it becomes liquid, and liquid hydrogen is stored and transported at minus 423° F. These fuels were earlier used on the Saturn I sites. In that case the liquid hydrogen was stored in a 125,000-gal storage tank. This was 38-ft diameter, double-walled and insulated by vacuum and expanded perlite, a glassy volcanic rock. Liquid oxygen was stored in a similar 125,000-gal tank with an outside diameter of 41¼ ft. The 4 ft separation between the inner and outer spherical tanks was filled with expanded perlite and pressurised with gaseous nitrogen. Smaller replenishing tanks for topping up were also provided.

The facilities for Complex 39 are very similar, except that the tanks are appreciably bigger, containing 900,000 gal of liquid oxygen and 850,000 gal of liquid hydrogen.

For the liquid oxygen a centrifugal pump with a discharge pressure of 320 lbf/in^2 pumps the liquid to the launch vehicle

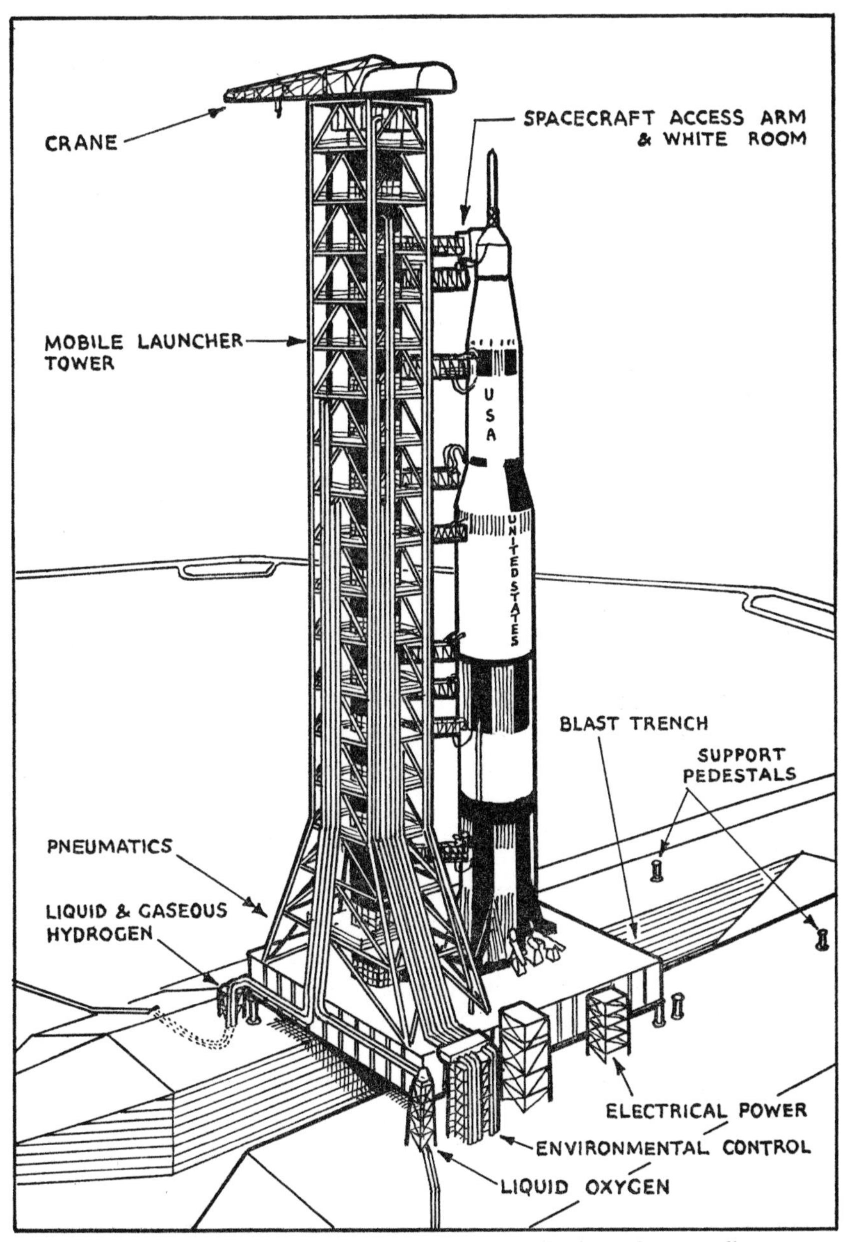

Fig. 34. View of Saturn V vehicle and Mobile Launcher standing over the flame trench of the Launch Pad hardstand.

at flow rates as high as 10,000 gal/min. A similar flow rate is attained with the liquid hydrogen, but in this case instead of using a pump a vapourising heat exchanger pressurises the storage tank to 60 lbf/in^2, and this pushes the fluid through the pipes. These very cold fluids have to be well insulated if they are not to boil and produce high gas pressures, and in both cases they are pumped through about 1500 ft of 10-in diameter, vacuum-jacketed, invar pipes from the tanks, to the umbilical interface towers, across to the Mobile Launcher and subsequently across the large umbilical tower swing arms into the Saturn V vehicle's stage tanks.

The RP-1 fuel is less of a problem and is stored in three tanks each with a capacity of 86,000 gal, and pumped into the S-IC stage at 175 lbf/in^2 and 2000 gal/min.

The Mobile Launcher, of course, combines two functions. It acts as a launch platform from which the rocket vehicle is fired at the launch site, while in the Vehicle Assembly Building it serves as an assembly platform upon which the space vehicle is erected and checked out. During pre-launch operations at the pad, the Saturn V vehicle's needs are served via nine umbilical attachments supported at the extremes of individual swing arms, which reach across the space between the vehicle and the umbilical tower like so many tentacles. These are similar to the swing arms provided on the Complex 34 tower but there are more of them.

Of necessity, the Mobile Launcher has to remain within the envelope of the environment produced by the rocket's engines at blast-off. Accordingly, the structure has been designed to withstand the extremes of temperature, stress and vibration generated within this envelope. The weight of the entire structure is close to 6000 tons, while it rises some 446 ft above ground level.

Although the Mobile Launcher consists of two major structural components–the launcher base and the umbilical tower–both are permanently secured together, forming a complete structural unit. They are discussed separately below simply for ease of presentation.

The launcher base is an all-welded steel structure of rectangular, box-like form, being 25 ft deep, 160 ft long and

135 ft wide. Internally the base consists of two levels designated A and B, which are compartmented. There are twenty-one compartments on level A and twenty-two on level B, each of which contains a variety of operational test and launch equipment vital to vehicle check-out procedures in the Vehicle Assembly Building and operations at the Launch Pad. This equipment includes computer systems, digitally controlled equipment for loading propellents aboard the Saturn V stages, hydraulic test sets and charging units, systems piping and ducting, instrumentation racks, electrical power systems and distributors, propellent and pneumatic lines, and so on.

Each compartment is either air-conditioned or ventilated as required. The entire base is also equipped with communications systems and the all-important fire protection system, which operates during vehicle lift-off and provides a deluge of water against the effects of rocket-engine blast. The floors within the base are shock-mounted and critical items of equipment are spring-mounted so that they receive less than $\frac{1}{2}$ *g* accelerations from mechanically induced vibrations between 0·1 and 1000 cycles/second. Acoustical isolation is also provided to reduce the noise level generated by the five F-1 Rocketdyne engines of the S-IC first stage to 110 decibels in the computer compartments and 130 decibels in other compartments.

A 45-ft square opening, extending the full depth of the launcher base, provides an exit for the F-1 engine-exhaust flames, which are distributed equally across the leading edge of the huge wedge-shaped flame deflector located on rails in the flame trench, the deflector being situated directly below the centre-line of the space vehicle. This square opening is lined with replaceable steel blast shields, which are independent of the main structure, and which are cooled by a water curtain initiated 2 seconds after lift-off.

A Hold-down Arm is located centrally along each side of the square exhaust opening. These devices are installed on the top deck of the launcher base and are situated on special strong-point support structures which are integral with the base structure. These arms provide the physical support for the vertically mounted Saturn V vehicle. They not only secure the rocket while in its passive state but they restrain it after engine ignition for about 7 seconds until the full engine thrust has been developed. When thrust is assured, the arms automatically release the vehicle for flight.

Three Tail Service Masts are also located on the top deck of the launcher base. They are situated in a 'Y' configuration about the periphery of the S-IC stage base between the Hold-down Arms. The Tail Service Masts support service lines, which provide fuel loading, air-conditioning and emergency liquid oxygen drain-off facilities for the rocket vehicle first stage. Umbilical carriers attach these service lines to the stage and they are pneumatically separated from the launch vehicle when approximately 3 in of lift-off has been achieved. The Masts have the capability of releasing, retracting and protecting the service lines from engine heat and blast effects.

Access to the launcher base interior is obtained through five personnel doors, and there are sixteen equipment access hatches located on the top deck. One door affords access to Level B from the Crawler-transporter only during the mobile transfer phase between the Vehicle Assembly Building and the launch site. All the doors are of the marine bulkhead type and are ringed with locking levers–indeed, the whole construction of the launcher base is reminiscent of ship building practice.

The umbilical tower is a 380-ft-high steel structure which is permanently positioned on the launcher base. It provides support for the Apollo Command Module access arm and eight swing arms for direct access to the launch vehicle stages. A total of ten umbilical carriers are attached to seven of the service swing arms and an environmentally conditioned chamber–the White Room–is supported on the Command Module access arm. This is attached to the Apollo spacecraft for astronaut entry and exit.

The tower also provides seventeen work platforms at varying levels, together with distribution facilities for propellents, pneumatic (gaseous nitrogen, helium and air), electrical, hydraulic and instrumentation systems. Other systems, such as closed-circuit television, communications, water and environmental control are also routed up the tower for use at various levels. The first two platforms are 30 ft apart, while the remainder are 20 ft apart. The working area of each platform is a square of 40-ft sides.

The distance between the vertical centre-line of the tower and the launch vehicle to 80 ft. Theoretical drift curves for the ascending rocket suggested that this distance should allow a clearance of about 15 ft for the worst conditions at lift-off. Some 29 seconds elapse from lift-off before the vehicle clears the top of the umbilical tower.

Mounted on top of the tower structure is a hammer-head crane with a hook extension of 468 ft so that it can reach, if necessary, right down to the bottom of the flame trench. Remote control of this crane is provided through portable plug-in-type consoles from any tower level, including the launcher deck.

Two high-speed electric lifts are centrally located within the tower structure. Capable of travelling at 600 ft/min with a load of 2500 lb, they serve eighteen landings between the upper compartment level within the launcher-base interior–Level A–and the 340-ft level of the tower. A low-pressure nitrogen 'blanket' system envelopes the electrical system in each car to avoid spark hazards–very important when so much explosive material is being piped aboard the Saturn V for fuel. In addition to normal freight and passenger service, these lifts also serve as part of the astronaut escape system. In the event of an emergency, the astronauts evacuate the spacecraft, traverse the access arm, take a speedy ride into the launcher base via the lifts and slide down an escape chute to a safety compartment within the concrete structure of the launch site. This is similar to the Complex 34 arrangements and, as there, an escape slide-wire has also been fitted.

Each of the nine swing arms, which extend from the umbilical tower, is supported at the top and bottom by a massive tower-mounted hinge assembly. These arms provide launch vehicle and spacecraft access for personnel and support the umbilical service lines required to sustain the vehicle at the launch site. The service lines consist of both rigid and flexible pipes of various sizes and materials. They interface with the piping installed on the tower and carry the required fuel and oxidiser (both loading and venting), pneumatic services, air-conditioning, purge, electrical power and monitoring connections to each umbilical carrier located at the vehicle end of seven of the swing arms.

Two arms do not have carriers. One of these serves as a service platform, and the other is the access arm to the Apollo spacecraft.

The Swing Arms separate from the space vehicle at varying time intervals during the last phases of countdown and lift-off, and they retract against the face of the umbilical tower in a matter of 6–8 seconds. These arms are interesting and complex mechanisms in their own right and will be introduced again later.

Three Mobile Launchers were originally built for Complex 39 and when not in use they are positioned in a parking area just north of the Vehicle Assembly Building. Whether inside this huge building, on the Launch Pad or in the parking area, each Mobile Launcher is seated on six pedestals about 22½ ft high; this gives sufficient headroom for the Crawler-transporter to drive underneath for transporting the Mobile Launcher to another site. A total of six pedestals, each 3 ft in diameter, are used to form a Mount Mechanism System at each site. This system is designed to withstand horizontal loading due to expansion and contraction and a variety of wind conditions through three arrangements of extensible struts disposed diagonally against four of the six pedestals. Each pedestal carries its share of the vertical loading and has a height adjustment of plus or minus 2 in. Before the Mobile Launcher is lowered on to the Mount Mechanism System by the Crawler-transporter, each column is adjusted by rotating the huge bolt collars until all pedestals are of the correct height with respect to the launcher. The Mobile Launcher is then connected to each pedestal by twelve massive high-strength bolts.

An additional support system is used when the Mobile Launcher is located at the launch site. Here the Saturn V vehicle is fuelled up so that the launcher base has to carry the full 3000-ton load; it is also necessary to make the launch platform more rigid for the actual lift-off sequence. Consequently, four additional extensible columns are positioned under the launch platform, one at each corner of the engine exhaust opening. Each column has an 18-in diameter double-acting hydraulic cylinder with a 22-in stroke, operating at a pressure of 4000 lbf/in^2. Load is transmitted through a spherical bearing in the base of each column, which also provides a means of correcting any vertical misalignment. After subsequent positioning and RP-1 propellent loading, hydraulic pressure is applied to establish a pre-load between each column and the launcher base. This load is maintained by the use of wedges, which effect a mechanical lock. After locking, of course, the hydraulic pressure can be removed.

The Mobile Launcher arrives at one of these sets of Mount Pedestals aboard the Crawler-transporter, which supports the combined weight of the launcher and the space vehicle on four flat-machined interface pads. The Mobile Launcher has four square projections which complement the positions of the interface points on the Crawler-transporter, and the latter must be positioned under the launcher within 2 in in both lateral and longitudinal directions, this being the maximum acceptable misalignment.

After the chassis of the Crawler-transporter has been 'jacked' into position and the four interface pads meet the projections on the underside of the launcher base, the entire load is secured in position by eight powerful air-driven motor screws arranged at each interface point on the roof of the Crawler-transporter chassis. Although the weight of the Mobile Launcher is enough to keep it in position in normal conditions these locking screws are additional insurance against wind loads or emergencies.

When the Mobile Launcher is in use, electrical power is required through interfaces for different conditions of service. One source must provide for motor-driven equipment and lighting, another for electronic systems which are sensitive to voltage variations in the power line. Indeed, four separate services are provided.

Instrumentation power is supplied from a sub-station to a transformer on level A of the launcher base, from which power is distributed through instrumentation and d.c. static power panels, when it will be used to monitor critical launch-vehicle control, safety and instrumentation circuits.

Industrial power is also supplied via a transformer on level A and distribution is to power and lighting panels, two 400-cycle generators and the engine gimballing unit controller which governs the attitude of four of the five F-1 engines on the first rocket stage.

In-transit power as its name implies is power supplied from the Crawler-transporter to the Mobile Launcher when the latter is being moved from the Vehicle Assembly Building out to the Launch Pad. This power is supplied from two 250-hp diesels each driving a 150-kV generator within the Crawler-transporter chassis. This is a small supply and it is used for such services as limited air-conditioning, principally for the computers on board, threshold lighting in the launcher-base compartments, obstruction lights, lifts and so forth.

Emergency power is supplied by diesel-driven generators through fixed, shore interfaces at the VAB and Launch Pad.

Full environmental protection is provided, in the form of air-conditioning and ventilation, for all equipment during all phases of operation within the VAB or at the launch site. Only minimal environmental air-conditioning and humidity control is provided during transfer between these two points.

Although many other services–such as those for everyday drinking water and sewage–have to be provided from ground supplies to the Mobile Launcher, or vice versa, only one other system will be mentioned here.

Industrial water is provided at the Launch Pad for the all-important purposes of cooling the Mobile Launcher structure and its accessories during those seconds when it will be within the flame and blast envelope of the Saturn V rocket engines as they build up full power and lift-off. This water supply feeds facilities for soaking the launcher deck and launch tower and is fed through a pipe of 3-ft diameter at 50,000 gal/min, dispersion being accomplished through twenty-nine deck nozzles and twelve perforated pipes.

Water-fogging nozzles are situated on each of four tower levels on which propellent transfer sleds are situated. These particular nozzles provide a 90-degree-spray pattern directed towards the centre of the tower level. Perforated pipes are also mounted vertically to the inside walls of the engine exhaust chamber and several large nozzles give perimeter coverage around the chamber. Deluge capability varies from 125 to 250 gal/min depending upon the location of the water outlet. Each service swing arm up the face of the umbilical tower has a water-fogging nozzle system mounted above the arm, so that each arm is completely engulfed following vehicle lift-off.

These sprays and deluges are not normally seen by onlookers at the time of a vehicle launch, because the water turns to steam and is taken out with the exhaust flames and gases; however, when tested on an empty Mobile Launcher, they are quite a sight.

The other piece of heavy equipment moved about by the Crawler-transporter as part of the mobile concept is the Mobile Service Structure. Like most things at KSC this is very large, being 402 ft tall and weighing around 5000 tons. However, it is a fairly straightforward structure and a very short description will suffice. It is a 40-storey steel-trussed tower used to service the Apollo launch vehicle and spacecraft while they are on the launch pad, and it provides all-round platform access. Five platforms are provided and two lifts carry personnel and equipment to the various work levels.

The Mobile Service Structure is mounted at the Launch Pad, and at its parking position, on four Mount Mechanism pedestals, similar to those used for the Mobile Launcher, many of the services also being similar. It is removed from the Launch Pad 11 hours before lift-off–at T−11 hr as it appears in the Countdown programme.

12
The Vehicle Assembly Building

THE Vehicle Assembly Building is the largest and most impressive structure at KSC, not so much for its architecture–though it has a unique elegance of its own–as for its sheer massive size. It is 716 ft long, 513 ft wide and 524 ft tall, which is just a few feet higher than the revolving restaurant of the Post Office Tower in London. It has a floor area of 343,500 ft^2 and an internal volume of close on 130 million ft^3, which must make it one of the largest buildings in the world. Within this building the most intricate combinations of space modules and launch vehicles ever conceived by man are received, assembled and checked out to an advanced state of launch readiness, and one of the major contributions of this huge building to the overall concept of Complex 39 is the application of assembly-line techniques–on a rather gigantic scale–to what was previously 'one-off' work.

The VAB is really two buildings joined together, known as the High Bay, which is 524 ft high, and the Low Bay, which is only 210 ft high.

The main building is essentially a box-shaped configuration. This is not the best type of structure for shedding wind, but it is sound for minimising side-sway. It has been estimated that strong gusts of wind of up to 125 miles/hour will cause the building to sway as much as 12 in–and being in Florida, hurricane-force winds like this are to be expected sometimes.

In keeping with the flexibility inherent in the mobile concept, the structural framework and foundations of the VAB have been designed with a view to future expansion. At present, three of the Saturn V class vehicles can be erected simultaneously in assembly cells in the High Bay area. A fourth cell is available for later expansion.

Early layouts of this building revolved around an 'in-line'

concept, with the four cells being placed in a row, resulting in a long, narrow, though high, structure. However, in the quest for maximum operational efficiency, NASA engineers eventually established that four high bays placed in pairs located back-to-back and having a single transfer aisle common to

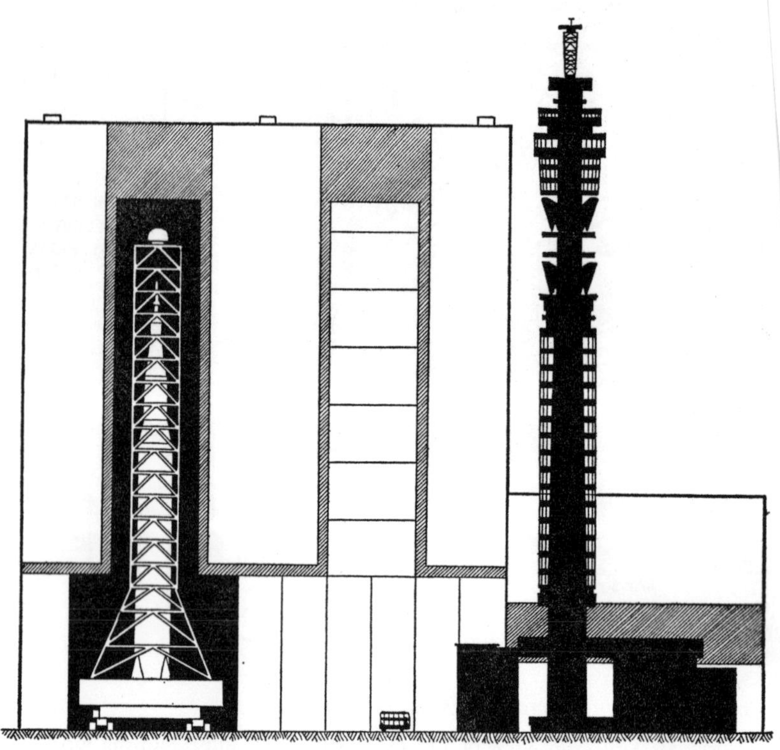

Fig. 35. The Vehicle Assembly Building 524 ft high, with a silhouette of the London Post Office Tower 620 ft high, for comparison. Also shown are a Mobile Launcher on a Crawler-transporter entering one of the Bays, and a London double-decker-bus for scale.

all bays would offer the best layout. This square configuration permits each assembly cell to be served by its own door, which opens on to the specially prepared Crawlerway over which the Saturn V will be transported to the launch site. Such a layout also eliminates the need for special high-

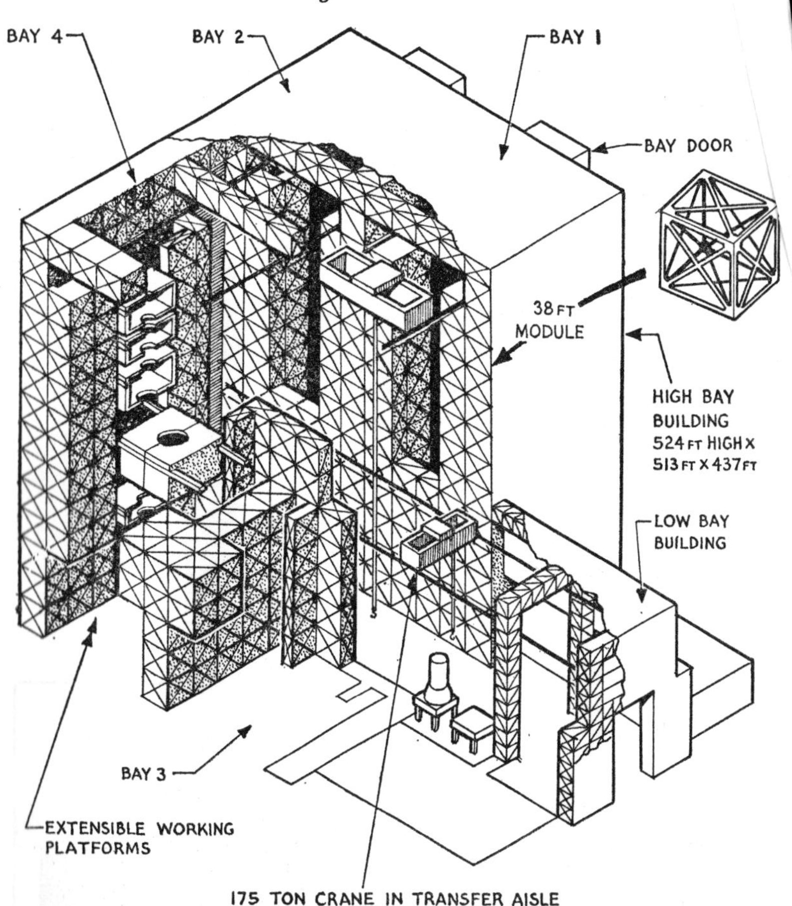

Fig. 36. Cut-away view of Vehicle Assembly Building showing form of construction.

capacity cranes in each assembly bay and the overall structure is more suited to taking the anticipated high wind loadings. The four immense door mechanisms are designed to operate in wind gusts up to 63 miles/hour, but only one door can remain open during periods of high wind.

The frames for the High Bay sections of the VAB have been designed and assembled in 38-ft modules in both the horizontal and vertical directions; this arrangement facilitated

optimum efficiency in detailing, fabrication and field erection. These frames have also been designed as an integral part of both wall and roof structures. This enables them to work in conjunction with each other as a composite structure of steel and concrete that will transmit horizontal wind loads more effectively and uniformly to the foundations. The roof consists of precast slabs joined by concrete grouts. Indeed, the entire design achieves effectively a monolithic type of construction through the unifying of walls, roof and structural framing.

The layout of the cells in the High Bay area is shown in Fig. 36. To assemble the huge 363-ft-tall Saturn V launch vehicle together with the Apollo spacecraft and secure it to the 440-ft-high Mobile Launcher, the VAB High Bay area is equipped with two 250-ton bridge cranes, each weighing some 500 tons with its 250-hp drive. One crane serves two High Bay assembly cells located on opposite sides of the transfer aisle and runs on rails located in the top trusses of the building.

Each High Bay assembly cell had to be designed so that the structural framework trusses were arranged in a manner that would not interfere with rocket vehicle erection. At the same time, future expansion capability had to be considered. Framework design had to be such that no member should be so vital that it could not be omitted by redirecting the stress-flow during facilities expansion. As a result, each vehicle erection cell within the High Bay structure is U-shaped, the base of which is called a diaphragm. Each diaphragm is slotted and this allows transfer of the launch vehicle stages from the centre aisle to the assembly position in each cell, by hoisting them 190 ft over the cell or bay framing situated between the aisle and cell. This system, known as a space-truss system, combines optimum stiffness with flexibility of layout.

Each High Bay assembly cell has five levels of extensible working platforms designed into the structure, these platforms ranging from one to three stories high. They are totally enclosed, self-contained structures which move out from the building structure and clamp round the launch vehicle in two halves. These platforms are thus virtually sizeable rooms fitting the rocket vehicle, and which are air-conditioned to provide comfortable working conditions for personnel.

Not all the assembly and check-out operations are carried out in the High Bay area. In the adjoining Low Bay area it is possible to prepare and check eight of the Saturn V upper stage elements, that is the S-II second stages and S-IVB third stages. Test areas, engineering support offices and laboratory space are also to be found in this area. A more conventional form of structural steel beams and girders has been employed throughout the design of this smaller portion of the VAB, which is 272 ft long and 437 ft wide–the length being considered as the direction in which the 175-ton capacity overhead bridge crane runs on into the High Bay area.

Whenever an S-IC first stage is received at the dock basin at Launch Complex 39, it enters the Low Bay area on transporter dollies, and it is partly checked out while lying in the horizontal position. The stage is then towed into the transfer aisle of the High Bay area where both the 175-ton crane and one of the 250-ton cranes rotate it into the vertical position. Following this manoeuvre, the 250-ton crane lifts the stage up and over the 190-ft-high truss towers and places it in the upright position on the deck of a Mobile Launcher, which has previously been positioned in one of the High Bay assembly cells.

The S-II second and S-IVB third stages are similarly received in the Low Bay area, together with all related accessories and equipment. These are handled by the 175-ton crane, which rotates them into the vertical position and places them on to individual servicing platforms running on rails set into the floor. From here the platform is backed into its respective check-out cell, from which point the stage, once assembled and checked out, is moved into the High Bay transfer aisle, where it is hoisted in the manner of the first stage and matched up with the preceding stage aboard the Mobile Launcher.

Both the High Bay and Low Bay areas of the VAB are provided with huge aeroplane hangar-type doors, which slide on tracks in the horizontal plane. However, above these sliding doors in the High Bay area, vertical sliding doors are fitted, the entire door height reaching 456 ft above ground level. The upper doors are in seven sections, each section weighing from 32 to 73 tons, and telescoping behind the next one in the vertical plane. The lower hangar-type doors are 156 ft across, while the upper vertical sliding doors are 76 ft wide. There are four of these huge door assemblies in the High Bay area, two on each of two opposite sides, one to each assembly cell. These doors are necessary to protect the space vehicles and personnel from the general vagaries of the climate, including hurricane-force winds. They are also needed to dampen the vibrational and pressure effects of sound-shock waves resulting from Saturn V lift-offs, for although these Launch Pads are 3 miles or so away it was estimated during the design stage that the noise level would be somewhere between 120 and 155 decibels.

This tremendous building provides a controlled atmospheric environment for selected regions about the Saturn V launch vehicles. This is vital to certain items of equipment during the assembly and check-out phases of the vehicle. Air-conditioning is provided by a 10,000-ton capacity unit, which is sufficient to control the temperature and humidity in a structure the size of the Empire State Building, or in 3000 average size houses.

It would be possible to control the temperature in the entire building, but it was thought inadvisable to do this in a building with so much unobstructed headspace in the upper regions as, whenever the huge doors were opened, pressure and temperature differentials could cause the formation of clouds near the roof and precipitation would ensue. Artificial rain, as it were, is most undesirable, of course, and to eliminate any such possibility, huge fans are at work above the 520-ft level within the building to keep the air in constant motion.

While the general form and layout of the Vehicle Assembly Building can be described, words are quite inadequate to communicate what it is like to be in a building where the ceiling above you is higher than the revolving restaurant on the Post Office Tower in London, or would tower 40 ft over the great Pyramid of Gizeh. Although the VAB is largely an enclosed space, over 60,000 tons of structural steel were used in the skeletal structure alone, and it took more than 2 million ft^2 of aluminium panels to cover the structural skeleton.

However, as we saw earlier, the site for KSC is of a very swampy nature–if it were not, the site would probably not have been available. Indeed, bad soil conditions are one of the penalties one pays for isolation. The result, of course, was that satisfactory high-capacity bearing qualities were not to be found in the soil at near surface depths. It was necessary, in fact, to sink steel pilings about 160 ft in order to gain the support of the more solid bedrock of what is known as the Ocala limestone formation.

Actual pile driving was begun in August 1963. As it was found that H-type pilings bent when driven through the upper rock strata, open steel pilings were used. Some 4300 piles were ultimately driven to form the foundation for this huge building. These piles are 16 in in diameter with $\frac{3}{8}$-in-thick walls, the accumulative weight coming to about 21,000 tons, and the accumulative length about 130 miles. After setting and driving, the piles were capped with concrete to a depth of 10 ft.

Earlier investigation of the sub-soil revealed that it would allow the application of a procedure known as 'sonic pile-driving' during the first 100 ft of depth. The sonic units produce longitudinal force impulses at relatively high rates–in the region of 100 cycles/second. This greatly accelerated the driving process in the early phases. The last 50–60 ft required to complete the pile-driving was accomplished through conventional diesel and steam drop-hammers.

Following the driving operation, every tenth pile was test-checked with thirty hammer blows to determine whether or not the safe loading value of 50 tons/ft^2 for each pile was assured. After pouring the concrete caps to form foundation sills, adjacent pile caps were interconnected by reinforced concrete tie-beams to assist the foundations to resist the large bending moments which would be transmitted from the superstructure. This superstructure, indeed, required some of the largest steel sections ever known to have been rolled at the time, the largest of these weighing 1305 lb/linear ft run.

For this building, and for KSC as a whole, a considerable amount of the work has been the result of joint planning between many government agencies and companies, notably between NASA and the US Army Corps of Engineers, who joined hands in the design, development and supervision of construction of the facilities at the Complex 39 site. Actual construction and much of the design work was delegated to private industry under contract, but this was still administered and monitored by the Corps of Engineers acting as agents for NASA/KSC.

And before we move on to other areas, just to highlight the level of work and effort put into this single part of the whole scheme, more than 100 miles of blueprints involving some 2500 separate engineering drawings were produced in the planning and design of this Vehicle Assembly Building.

13
The Crawler-transporter

As we have seen, the Crawler-transporter is one of the principal components of the launch support operation at Complex 39, for without it the mobile concept would not have been able to survive. It provides the means whereby huge items of ground-support equipment can be moved from one location to another. Its principal purpose is for moving the Apollo/Saturn V space vehicle and its Mobile Launcher from the Vehicle Assembly Building to one of the Launch Pads, 3–3½ miles away.

This operation is performed at the tortoise-like pace of 1 mile/hour. The maximum unloaded speed of the Transporter is 2 miles/hour, but going up the Launch Pad ramp with load the speed is reduced to only ½ mile/hour. Journalists sometimes refer to the Crawler-transporter as 'lumbering along at a clanking one mile per hour', implying that the operations are unduly ponderous and noisily performed. In fact, progress is so slow as almost to defy perception, and one is immediately struck by the quietness and smoothness of the entire operation. Such quietness is, perhaps, surprising in a monster of over 3000 tons carrying a load of 6000-odd tons, but the various systems are well muffled and controlled.

The entire moving operation must, of course, embody inherent stability, smoothness and steadiness in order to preserve the integrity of the many delicate systems aboard the passenger space vehicle. Hence the slow pace. Speed at this crucial stage just to save an hour or two would be quite unwarranted if it led to compromising the proper functioning of the many launch vehicle systems. A resulting failure could lead to hazarding a mission–which might have cost several millions of dollars and taken several years to set up.

As we saw earlier, there was some difficulty in selecting the type of design to start upon for a suitable transporter. After examining with no success ideas for employing a pneumatic-

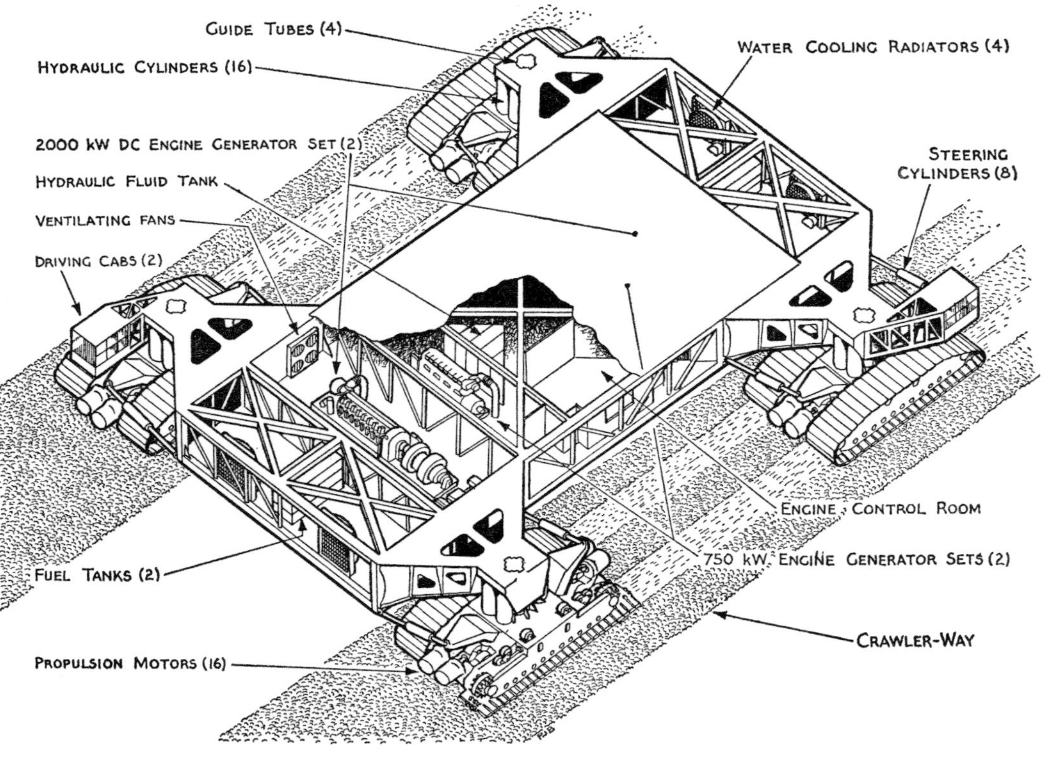

Fig. 37. The Crawler-transporter.

tyre transporter, a hovercraft machine, a rail system and a barge system–all of which would have taken a considerable time to develop–the latest of the Bucyrus-Erie Company's Shovels, 3850-B, came to notice. This Shovel–a gigantic earth shifter–was fully mobile and the load moved was close on 9000 tons, not far short of the load estimated by NASA engineers. This giant stripping shovel relied upon tank-track crawler principles to satisfy motivation requirements and hydraulic jacking for levelling purposes, methods which had proved successful in earlier and smaller pieces of equipment of this type since the early 1920s. It was evident that this machine represented the existing state-of-the-art in this area and it suggested reasonably low development costs and time. So the decision to design and use a Crawler-transporter as part of the mobile concept was taken on 25th July 1962.

The Transporter which emerged consists essentially of a Platform and Chassis supported at the corners through pillars by four crawler trucks, each having twin tracks as seen in Fig. 37. The depth of the Chassis is used to house all the necessary machinery and is very much like a ship's engine room; and a lot of machinery is required because, apart from the propulsion system, there are power and hydraulic systems needed for the Transporter and to sustain the Mobile Launcher and its rocket-vehicle load during transit. Also provided are a steering system and a braking system and, integrated with one another, there are also jacking, equalisation and levelling systems to keep the platform level and, thus, the Saturn V vehicle vertical.

The Chassis is of a rectangular shape, some 131 ft long and 114 ft wide overall. The frame is of rigid box-like steel construction capable of restraining and sustaining all normal operating forces and loads to which the machine is likely to be subjected in service. The Chassis is designed to operate as an independent unit, so that stresses induced by propulsion and steering are contained within the chassis framework and not transmitted to the transported items. The Chassis is a totally enclosed structure within which all equipment for operation is housed in accessible locations.

Ready accessibility is provided by hatches to facilitate

installation and removal of such items as engines, generators, pumps and motors. The interior is suitably illuminated and supplied with conditioned air. An enclosed area takes the form of a manned control room in which monitoring instruments, electronic equipment and other devices are housed. This room will accommodate several people during operations. Freedom of movement for personnel is provided in the form of visual inspection platforms, interconnecting catwalks and observation decks. There is, indeed, much more in common with the engine room of a ship than with a land-based vehicle. Access ladders are available to personnel, these being placed one on each side of the Transporter from ground level to the observation decks. The angle of these ladders is self-compensating so that the ground clearance remains uniform, a feature which, as will be seen, is clearly necessary.

The overall height of the Transporter is 20 ft, from the ground to the top deck upon which the Mobile Launcher rests for transporting. This deck is flat and about the size of three full-size tennis courts. Two operators' cabs–one at each end of the Chassis located diagonally opposite one another–provide totally enclosed stations from which all operating and control functions are co-ordinated. Adequate provision for air-conditioning, heating and illumination, together with considerable scope for movement, are incorporated for the comfort of operating personnel–necessary during the quite long periods of Transporter operation, for it takes many hours to complete the 3½-mile journey.

Apart from having at hand information concerning propulsion and steering, the cab operators are continually informed through displays of operating and attitude conditions throughout the journey. Communication between all manned work stations is provided in the form of headsets, boom-mounted, noise-cancelling microphones, paging speakers and amplifiers to provide, with additional equipment, an integrated system. In passing it can be noted that each cab driver, at least at one time, boasted what was thought to be the largest windshield wipers in the world–which is not surprising when one considers the general

Gulliver-like scale of everything in sight at the Kennedy Space Center.

Traction is obtained through four huge double-tracked Crawler Trucks, which support the Chassis at each of four corners, the vertical centre-lines of the trucks being located on a 90-ft square.

Deflection of the Chassis due to bending is kept to a minimum by locating the load-carrying interface points near the centres of the Crawler Trucks. Four massive guide tubes are permanently attached to the Chassis at each of the four centres. The Crawler Trucks pivot about these guide tubes, thereby facilitating changes of direction in Transporter operations. Each guide tube absorbs the shear forces produced when the Transporter is in service. The Trucks are capable of being steered in pairs at either front or rear, or of being centrally steered together permitting 'great-circle' or 'crab-like' changes in direction.

Manoeuvrability, without detrimental effects to the space vehicle on board, is one of the main advantages attributable to the crawler system, and the Crawler can, of course, be driven equally well from either end.

Each Crawler Truck, like some super battlefield tank, is provided with a pair of steel-link crawler tracks, placed parallel to one another either side of the guide tube. Each Truck is 24 ft wide overall; each link of the crawler tracks is $7\frac{1}{2}$ ft long and weighs about 1 ton.

The Crawler-transporter obtains its power from four 1000-kW d.c. heavy duty electric generators, located in the Chassis and driven in pairs by two 2750-hp diesel engines. These generators supply power to each of four electric propulsion motors mounted in pairs, fore and aft, on each Crawler Truck–16 in all. Two propulsion motors drive each crawler track belt through five-stage reduction gearing having a ratio of 168:1, and through the immense driving sprocket located at each end of the Crawler belt. The 750 rev/min of the propulsion motors thus produce 1 mile/hour on the track belt.

A disc-type braking system is located on the armature shaft of each of the sixteen propulsion motors. This system is capable of bringing the Transporter to a smooth gentle stop and is synchronised to operate so that power failure at one Crawler Truck will not result in accelerations beyond 2·6 ft/sec^2. Each motor is provided with a service brake and a parking brake. The service brake controls the movement of the Crawler-transporter and is hydraulically/air operated via a foot pedal in each operator's cab. The parking brake is designed to hold the Transporter while parked only; it is set by a double compression spring device and released by an air cylinder.

Manoeuvrability is a very important feature of this mechanical mammoth, when it is realised that it must position the Mobile Launcher with the Apollo/Saturn V vehicle within plus or minus 2 in at the Support Column interfaces. Positive steering provides this essential element of lateral and longitudinal accuracy. Steering is facilitated by huge hydraulic actuating cylinders, mounted in pairs, one at each end of each Crawler Truck. They operate on a push–pull basis, the ball bushing at one end being attached by pin connections that are integral with the Chassis, while the other end is attached to a massive steering lug, one being positioned at each end of each Crawler Truck.

The unique automatic levelling system built into the Crawler-transporter consists of sixteen hydraulic levelling cylinders–four on each Truck–together with an overall interconnecting hydraulic system, an automatic level-sensing system of the manometer type and a manual jacking system, the latter being installed as a back-up method in the event of a power failure or component malfunction. Four of these hydraulic cylinders are mounted around each of the guide tubes (about which each Crawler Truck pivots) and they provide the required vertical support for the Chassis. Levelling and jacking loads are transmitted to the Chassis through ball bushings at each end of the cylinders. Each cylinder has a total extensible travel of 6 ft and is capable of resisting the load resulting from a wind of 68 knots at 400-ft elevation up the tower–a force of about 300,000 lbf.

The Crawler-transporter takes on its load by being positioned under the Mobile Launcher or Mobile Service

Structure, and extending the jacks so that the Chassis is raised to a point whereby the specially designed self-centring support pads and locking systems engage. Sufficient jack travel must be available beyond this point to permit the transporter to 'level' while moving clear of the Mount Mechanism or Support Columns.

The levelling function is maintained by an automatic manometer level-sensing device which controls valves and pumps. This device causes fluid to be pumped to the low cylinder from the diagonally opposite cylinder on the other side of the Chassis. The levelling system is designed to maintain the Chassis within plus or minus 5 minutes of arc at all times and no point in the support plane is permitted to be out by more than 2 in at any time during operation. Limit switches are provided to prevent excessive extension or 'bottoming' of the cylinders.

This levelling facility is, of course, especially necessary with respect to the 5% ramp which the Transporter must climb to reach the hardstand of the Launch Pad. As might be expected, this is a tricky phase of operations and it is performed at a reduced speed and higher motor torque. Indeed, the Crawler-transporter has three speeds–2 mile/hour with low torque for the unloaded condition, 1 mile/hour with normal torque for the level loaded condition, and $\frac{1}{2}$ mile/hour with high torque for the ramp-climbing condition.

Two diesel engines within the Chassis drive two 750-kW 3-phase generators, each of which provides one half of the levelling, jacking and steering power requirements. Two smaller 150-kW generators supply the Mobile Launcher with the necessary power supply while it is being transported.

The Crawler Trucks of this huge mechanical transporter are placed at 90-ft centres, each Truck being some 24 ft in overall width. Each pair of crawler track belts has been designed to provide a substantially uniform bearing load over a minimum area of 500 ft^2 at 1 in depression of the resilient roadbed or Crawlerway. Idler rollers have been provided and are used so as to ensure this desired uniform loading over the length of the crawler track belts.

The two Launch Pads and the Vehicle Assembly Building are interconnected by this Crawlerway, which is a specially prepared roadbed. The Crawlerway consists really of two parallel roadways each 40 ft wide and spaced on 90-ft centres giving a total width of 130 ft.

This Crawlerway presented quite a number of design problems of its own. The loading came out to 12,000 lbf/ft^2 and nothing of this order had ever been carried over soil conditions such as those found at the Kennedy Space Center site. An extensive feasibility study was carried out to determine the materials and dimensions necessary to perfect such a roadbed. This study suggested that the Crawlerway had to be constructed in three layers and then sealed with a prime resilient coat. Several factors influenced final selection of the construction method and these included sub-surface soil conditions, the effects of high unit loading on the Crawler-transporter and its load, the amount of elastic and permanent surface deformation that could be tolerated following repeated use, and the effects of prolonged periods of parking such a load in one spot due to high winds or power system failure in the Transporter.

The basic Crawlerway is some 6 ft thick. The base layer consists of 2½ ft of selected hydraulic fill stabilised and compacted to 95% density, beneath which the bed is of existing earth or compacted sand as required, to provide the desired grading. The middle layer is some 3 ft thick and consists of compacted, crushed rock and soil cement base consolidated to 100% density. These sub-layers are then topped off with a prime coat of asphalt.

Two Crawler-transporters were built for operation at Complex 39 and they have proved very successful. However, quite a lot of new ground was being broken and, although the contract for these was given to a 'crawler shovel' company, there were quite a few 'teething troubles' in the early operational stages. The first Crawler-transporter was delivered on 23rd January 1965 and right from the start the automatic levelling system misbehaved. Eventually, the trouble was traced to the servo-mechanism system—with an amplification of 2000:1 in the circuitry it proved too sensitive and continually over-corrected errors in levelling.

In July of that year a test with a load of nearly 8 million lb led to failures in a number of the roller bearings which transmitted the load from the Crawler Chassis to the track treads. These failures were attributed to the fact that with the then Crawlerway construction resiliency did not average out loads on the bearings sufficiently and some were obliged to take peak loads very much above that for which they were designed. Eventually they were replaced by new sleeve and thrust bearings instead of with heavier grade roller bearings.

Surface resiliency is very necessary when it is realised that the transport beams for the idler wheels on the Crawler Trucks which run on the track-belt links may deflect as much as $\frac{1}{2}$ in under load. These deflections, together with any irregularity in the Crawlerway surface, would serve to intensify stress concentrations within the Transporter structure if the road surface were rigid.

A resilient surface is also desirable to minimise the possibility of increased stresses at the curves and to produce a sliding surface in order to reduce friction while negotiating such curves.

Indeed, steering proved more difficult than expected. While on the Bucyrus-Erie stripping shovel the coefficient of friction was found to be 0·3, for the Crawler-transporter it was estimated to be 0·4, and the vehicle was designed with an assumed value of 0·6. In the July run, the actual coefficient of friction turned out to be higher than 1·0. As a result, not only were the stronger bearings fitted but a number of improvements were introduced to the Crawlerway surface composition, the entire Crawlerway being topped with 4 in of river rock on the straights and 8 in on the curves, thereby producing a more resilient surface which reduced steering friction. The Crawlerway tracks are now 'swept' and levelled off after each Crawler pass.

The two Crawler-transporters eventually cost about \$14$\frac{1}{2}$ million or £6 million and an additional \$1 million was spent on the Crawlerway changes. As Dr H. Kurt Debus, Director of Kennedy Space Center said, 'We are not buying a car off the line; the Crawler is a development job and there are no handbooks to turn to.' It was, he pointed out, quite impossible to simulate the very large loadings without costly and time-consuming full-scale models, and a certain amount of 'trial and error' was, in this case, the best route.

14
Launching Equipment

WE are now going to look briefly at three important items on the Mobile Launcher–items which are, indeed, related to one another.

Consider first the vast Saturn V launch vehicle standing on the Mobile Launcher over its exhaust chamber. It has got to

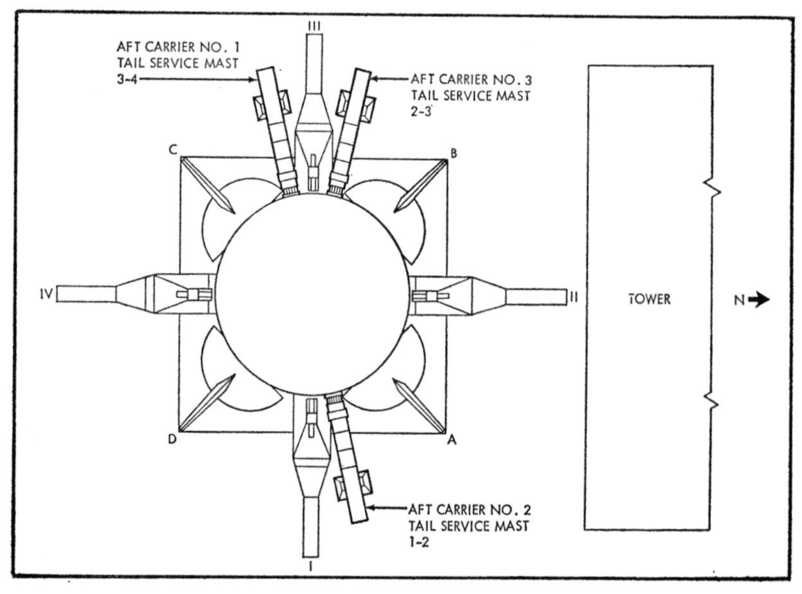

Fig. 38. Plan of Mobile Launcher Deck.

be firmly secured so that it is quite rigid during assembly, during transportation to the launch site, and during its stay on the Launch Pad in all weathers. It has got to be held down after engine ignition until all engines register full thrust. Then, quite suddenly, it has to be free for lift-off.

The rocket vehicle, weighing nearly 3000 tons when fuelled up, stands on four Hold-down Arm structures, one to

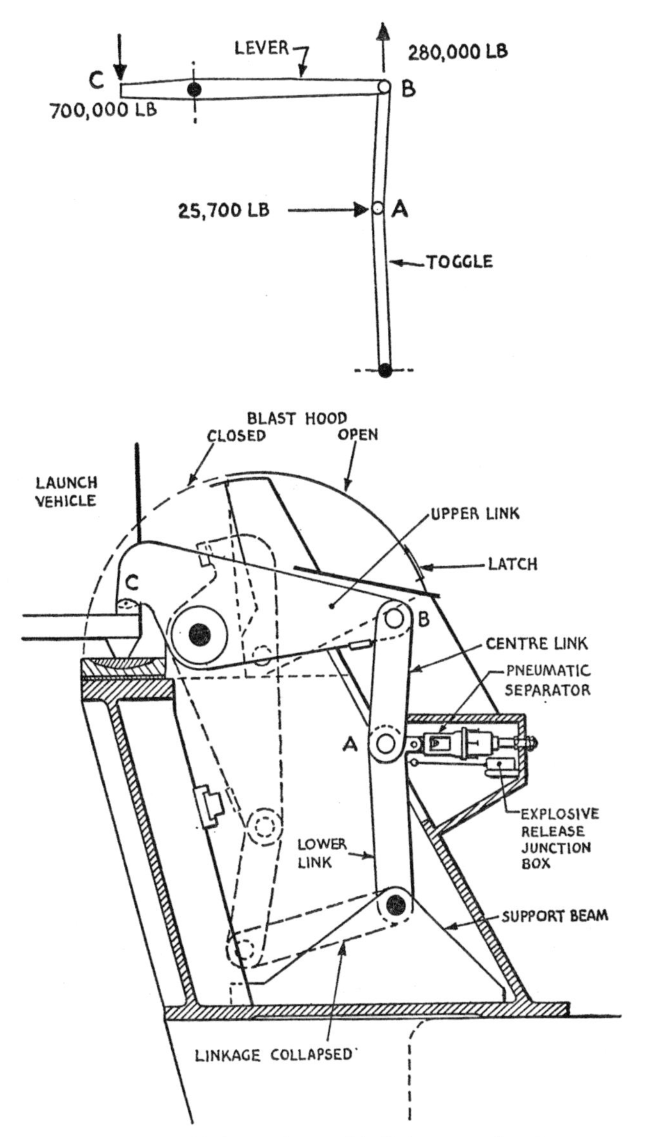

Fig. 39. Hold-down Arm with linkage mechanism.

each side of the blast chamber opening in the Mobile Launcher deck, as seen in Fig. 38.

The S-IC first stage rocket base ring is clamped at each Arm between an adjustable head and an upper link with a gripping force of 700,000 lbf, which is achieved through a toggle linkage. As shown in the inset of Fig. 39, a small force applied at A will create a large force at B, and the nearer the toggle gets to straightening out, the greater the multiplication factor. This principle is used, for the Hold-down Arms, where the force applied at A is multiplied at B and, by moments, is increased still further at C. This force is applied by an hydraulic cylinder connected indirectly at A, the links, once pre-loaded, being held in place mechanically by a separator link, so that the hydraulic pressure can be released and the cylinder removed. When this separation link is severed, the whole Hold-down Arm linkage collapses under its own weight to the position shown by broken lines, and the launch vehicle is free.

However, the 'buck has been passed', as it were, to the separation link, which now has to take the load safely and securely but which at a signal must instantly part. There are a number of ways of doing this, but here this task is carried out pneumatically by incorporating a pneumatic separator into the separator link.

This separation link can be seen in Fig. 39 and is shown in section in Fig. 40. The whole unit is held together in one piece by eight balls in the ball lock. At the time of rocket vehicle release, pneumatic pressure enters the cylinder and moves the piston to the left against the force of four springs, which up to then have prevented premature release. The piston moves over until an annular groove reaches the eight balls, which can then retract inwards, freeing the outer sleeve. This is shown in the illustration as Pneumatic Release and the whole operation takes place in a fraction of a second. As a 'back-up' a redundant Explosive Nut Release is also built in, although this only operates if the primary mode fails. When the Hold-down Arm links begin to collapse a lanyard on them disengages the explosive firing circuit on that Arm. If, however, through some fault the ball lock does not operate, this firing circuit remains intact and 0·19 sec after the pneumatic release signal, another signal is generated to fire the nut, which then explodes and releases the linkage.

Bearing in mind the possibly disastrous consequences which might result from some unforeseen failure, certain pre-

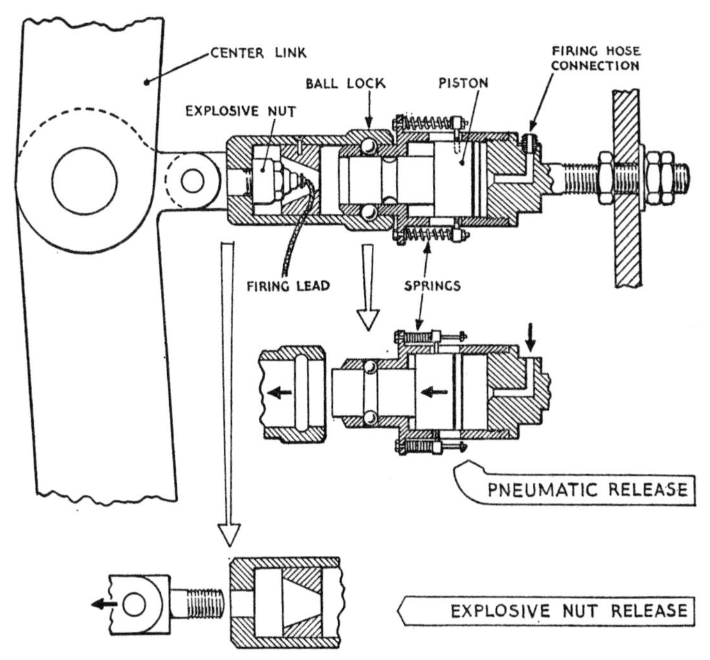

Fig. 40. Section through Separation Link.

cautionary measures have been built into the release system. The whole mechanism of release depends upon there being helium gas at 1500 lbf/in^2 available to overcome the spring pressure in the pneumatic separators. A valve is opened just before firing which pressurises the circuit to this level, and the attainment of this pressure closes a pressure switch, this in turn closing the engine ignition arming circuit. It is the making of this last circuit which allows the F-1 engines to be ignited at T$-$7 sec. If sufficient helium gas pressure is not available, the rocket vehicle cannot be fired.

The vehicle is restrained until thrust builds up to $7\frac{1}{2}$ million lbf at zero time, when a 'release' signal is generated and this opens valves which energise the plungers in each of the four ball-lock mechanisms. Just in case, in these 7 seconds, there is a decay in the helium pressure, a check valve able to sense this can connect up a supplementary supply of helium if necessary.

These Hold-down Arm Assemblies are naturally quite large and heavy. The base of each is 6 ft 4 in wide by 9 ft 9 in long, and they stand some 11 ft high, and weigh over 20 tons each. The centre and lower links of the toggle mechanism weigh about 800 lb each, and the massive top link reaches 2600 lb, so that a winch mechanism has to be incorporated just to raise the linkage into position before the pre-loading operation can start.

Great care is necessary in releasing a monster rocket vehicle such as the Saturn V. As the Hold-down Arm linkages collapse almost instantaneously, the vehicle could well accelerate too rapidly during the first few inches of travel–similar to the 'kick' in the back one feels when a jet plane releases its brakes to go rushing down the runway on take-off –unless some further restraint is provided. The devices which offer this restraint, called Controlled Release Mechanisms, are in principle remarkably simple. They are based on the wire-drawing principle, whereby wire is drawn through a die to reduce its diameter. The Saturn V is held down by sixteen special bolts, called pins, four fixed to brackets on each of the Hold-down Arm bases. These pins are in the vertical position, so that the lift-off thrust acts along their axes. In order to rise, the rocket vehicle has to draw these pins through dies reducing the pin diameter from 1·795 to 1·563 in., the effort to do this providing a restraining force of around 1 million lbf. However, the part of each pin to be drawn through its die is not of constant section–it tapers until it becomes the same diameter as the die itself. Consequently, the force needed to pull the pins through the dies continually decreases until, at about 6 in of rise, it reaches zero and the vehicle is fully free. While this system is simple in principle, to work well it has to be ensured that all the draw rod-and-die dimensions, materials and heat treatment are identical within close limits.

Mounted externally to two of the Hold-down Arm Pedestal Bases are Lift-off Switch Assemblies. These are simply pivoted arms which, by turning a cam, activate plunger-type microswitches. Working co-operatively, these two assemblies close switches which firstly monitor hold-down release and vehicle lift-off–so that the controller knows when lift-off has actually occurred–and secondly, provide a signal to initiate retraction of the in-flight Service Swing Arms on the Mobile Launcher tower, which we shall look at briefly later.

The next items of interest are the Tail Service Masts. These take their name from the fact that with early rockets the pneumatic and liquid hoses, and the various cables which had to be connected until lift-off, were suspended from a mast-like structure. With the huge and complicated Saturn V vehicle, these Tail Service Masts are rather more sophisticated.

There are three of them, mounted on the launch deck between the Hold-down Arms (Fig. 38 and Fig. 41) and they take the form of 'see-saws' on trunnions. One mast has a 6-in liquid oxygen hose, another a similar RP-1 line, and all three carry masses of electrical cables and pneumatic service lines. These come up through the Mobile Launcher deck into the Mast Pedestals, round the trunnions and along the mast arm, whereupon they connect up with similar pipes and cables on the rocket vehicle stage, the interface in each case being made up of a 'flight carrier' and a 'ground carrier' on the vehicle and Mast respectively.

At lift-off, all the connections have to be broken, the hoses having self-sealing couplings so that fluids are retained; the masts have to be swivelled up into the vertical position well clear of the rocket's flight path, and the exposed flight and ground carriers have to be covered to protect them from the fiery blast of the five F-1 engines as they emerge from the square exhaust chamber in the launch deck. All this has to be accomplished in 3 seconds and with a very high degree of reliability, since a 'hang up' would seriously jeopardise the flight.

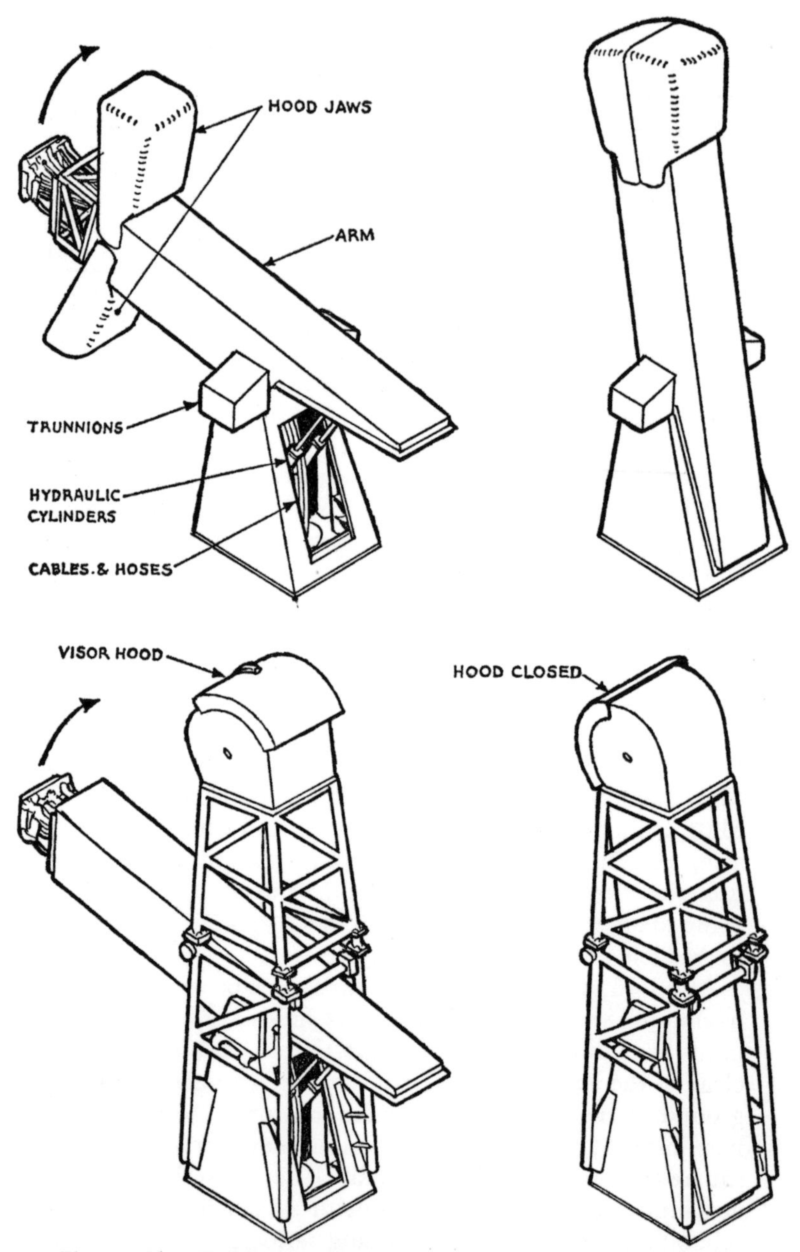

Fig. 41. *Top:* Tail Service Mast as used on first Saturn V launch, left supplying rocket stage and right retracted.
Below: Tail Service Mast as modified (see Chapter 15).

These Tail Service Masts look complicated, especially at the ground-carrier position, where all the hoses and leads are being carried to the rocket stage, as well as to the mechanisms for powering its own operation. In principle, however, they are simple, as must be so for reliable operation.

First a 'quick release' mechanism is required, though this must lock the two carriers together positively until the correct moment. The mechanism adopted is one with collapsing fingers, which are normally extended to lock with the flight carrier by a conically-ended rod. At the appropriate time, this rod is withdrawn, the fingers collapse inwards aided by a spring, and the two carriers are free of one another.

Next, the carriers have to be separated cleanly. This is done by having a push-off foot mounted round the above lock, and this foot extends as soon as the carriers are free, pushing the two carriers apart.

With respect to these unlock and disconnect operations, neither must fail and both must occur automatically at the right time. For both purposes the actual movement of the rocket vehicle as it lifts off is used. As the vehicle begins to rise, the ground carrier goes with it and, as the mast arm is mounted on trunnions, the angle of slope of the carrier arms start to change with respect to the ground-carrier face plate. This motion trips a pneumatic valve via a cam, and this collapses the fingers in the lock, while the withdrawal of the central rod itself trips another switch, which operates the push-off foot.

If either or both fail, both operations can be carried out mechanically. This time, instead of the rising carrier giving signals only for the disconnect and retract motions, other cams, being positioned by the changing angle of the carrier arms as the vehicle continues to rise, actually do the work of pulling out the locking mechanism rod and of pushing out feet for separation.

The act of separation operates a valve which supplies hydraulic pressure to two pistons to rotate the whole Mast to the vertical position. The rear end of the Mast is filled with lead; this more than counter-balances the front end so that, in the event of an hydraulic failure a Tail Service Mast would tend to rotate to this clear upright position of its own accord.

After rising 10 ft, the five F-1 engines emerge from the exhaust chamber in the launch deck, and the Hold-down Arms, the Tail Service Masts and other launch deck equipment are then within the intense heat and shock envelope of the rocket engines' exhaust. To afford protection to this equipment, these items have had to be made very strong and, on the Masts, the hoses and cables had to be completely covered at this time. There were only two places where they were exposed–at the rear of the base pedestal where the hydraulic rotation cylinders operated, and at the carrier itself. In the first instance, when the Mast reached the upright position, the rear, lead-filled end of the arm made a complete seal over the back of the pedestal base. In the second, two large hoods, like clam-shells, were provided in the original design, to snap shut over the carrier as it rose. Hydraulic actuators keyed to the trunnion shaft forced hydraulic fluid to slave cylinders mounted on the hood pivot shafts; in other words, hood closure was a function of Mast Arm rotation, but the connection was a local hydraulic one, rather than a mechanical one.

Throughout these designs there runs, of course, a general philosophy. The primary aim is that the flight of the launch vehicle shall not be jeopardised. The secondary aim is that the equipment itself shall be adequately protected so that it can be used again. Optimum power-controlled methods of carrying out all the necessary operations in sequence are provided and, in case of failure at any point, suitable back-up facilities are at hand. The back-up methods ensure that the primary aim is always met but, remembering that there is only 3 seconds for the entire operation, they cannot ensure in all circumstances that the secondary aim is always met. The 'clam-shell' hood has since been replaced by another, more simple design, as we shall see later.

The other important items to be discussed here are the Service Swing Arms mounted on the tower of the Mobile Launcher. Their functions are very much the same as those of the Tail Service Masts except that the service lines they carry are to the upper stages of the Saturn V vehicle and the Apollo Spacecraft. To a first approximation they may be considered as 'tail-service masts' up the tower and moving in an horizontal rather than a vertical plane.

There are nine such swing arms on the Mobile Launcher tower. One is the Command Module Service Arm; one is primarily a work platform to the S-II Aft stage, and the other

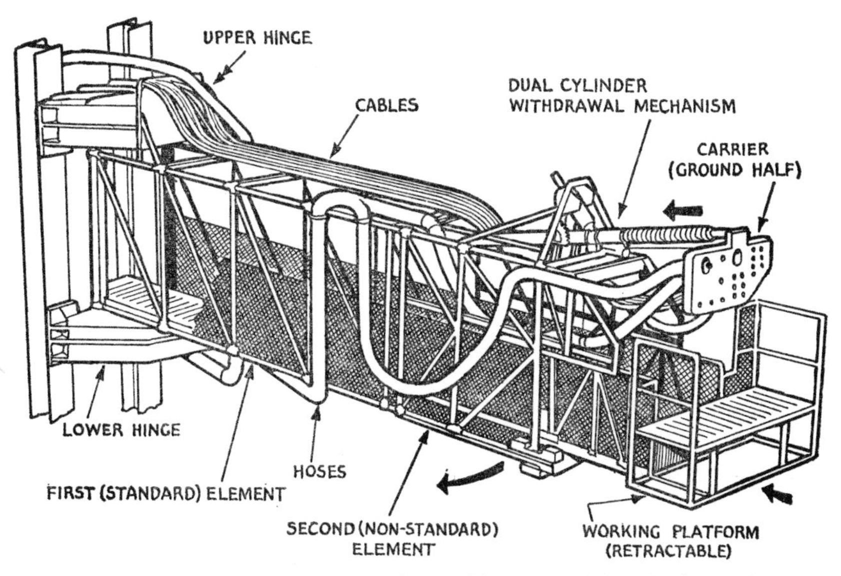

Fig. 42. Service Swing Arm. This one, No. 5, supplying the forward end of S-II stage, is typical of those fitted to the Umbilical Tower.

seven carry between them ten umbilical carriers and service lines required to sustain the vehicle at the launch pad. These are composed of both hard and flexible lines of various sizes and materials, carrying fuel and oxidiser loading and venting provisions, pneumatic system service connections, air-conditioning and purge, and electrical system powering and monitoring connections.

Excluding the Command Module Service Arm, which is a separate design concept, these Service Swing Arms may be divided into two groups. The S-II Aft Stage arm and the

S-IC Intertank and Forward arms are termed 'pre-flight' swing arms and these are retracted and locked against the tower face 15 seconds before lift-off (T−15 seconds), giving ample time for confirmation of correct operation before ignition at T−7 seconds. The others are 'in-flight' arms and they retract at lift-off after receiving a command signal from the Lift-off Switch Assemblies located on the Hold-down Arm bases, as mentioned earlier, and take 6–8 seconds for withdrawal and retraction.

Although the Service Swing Arms are all different, varying in length and with what is carried on each, they are all similar in principle and have parts in common. Indeed, the inboard half of each arm and the hinge attachments are the same, in general, for all the arms, only the outboard sections being tailor-made.

The general construction of a typical Service Swing Arm is shown in Fig. 42. The inboard end, nearly 23 ft long, is a Pratt truss with the upper and lower chords connected by members designed for load reversals. Upper and lower hinge plates attached to this part support the Arm in the tower hinges. The outboard ends vary, but they are basically a K-truss design of the length required, and most have an Extension Platform at the extremity which serves as a walkway to the Saturn V vehicle. These platforms are cleared and withdrawn about 6 hours before lift-off (T−6 hours).

At the end of each arm is an umbilical carrier plate–in some cases more than one–very similar to those in the Tail Service Masts; and, of course the problems here are much the same, i.e. those of reliably unlocking and withdrawing. A ball-lock, similar to those on the Hold-down Arms, is used to connect the flight and ground halves of the carrier, and unlocking and pushing off are carried out by power when the command signal arrives. Once again, there is a redundant, i.e. back-up system called a 'cam-off mechanism' which performs these functions after another 2 in of lift if the primary mechanism has failed.

As soon as the carrier is unlocked and ejected from the vehicle, an Umbilical Carrier-release Confirm Switch energises the service arm retraction and umbilical carrier withdrawal mechanisms. The latter is essentially a pneumatic and hydraulic cylinder unit lying along the axis of the service arm and mounted in a universal joint. The end is, of course, attached to the carrier which, by virtue of the universal joint, is allowed a considerable amount of movement relative to the arm itself. When commanded, it rapidly withdraws the ground carrierplate, while at the same time the Service Swing Arm has started to rotate about its hinges towards the tower, an hydraulic cylinder being fitted in each of the upper and lower hinge assemblies for this purpose. The arm speed is controlled by a cam-operated deceleration valve, which controls the discharge flow rates from the hydraulic cylinders. Linear shock absorbers are fitted on the tower; and the Service Swing Arm, once flush against the tower, is latched at its outboard end and supported on a specially provided support.

The latches and supports are more important than they might sound. Without them a Swing Arm could retract and bounce back inside 12 seconds–and it takes over 20 seconds for the Saturn V vehicle to clear the Mobile Launcher. The arms need extra support against the tower because as soon as they are secure they are subjected to the blast forces of the five F-1 engines passing over.

A back-up retraction system is, of course, mounted in the tower for each arm. This takes the form of a block-and-tackle system with the sets of pulleys pushed apart by an hydraulic cylinder. The cable, after being passed round the pulley sheaves, is led out of the tower to be fixed to the Swing Arm about 22 ft out from the hinge position. This system works simultaneously with the hydraulic cylinders operating directly on the hinges, but normally it takes no load, merely taking up cable slack. In the event of the hydraulic system to the hinges failing, however, this redundant cable retract system immediately takes over and completes the job.

There is nothing very exciting about these Swing Arms. They are 'bread-and-butter' jobs to the engineers, but they are nevertheless very important as is their correct functioning. At the time of lift-off, except for the astronauts at the top of the rocket vehicle, there is not a living soul within miles, so that these Service Swing Arms–and the Hold-down Arms, the Tail Service Masts and all the other things going on at the Launch Pad–have to be monitored remotely in the Launch Control Center.

15
Trials and Tribulations

THROUGHOUT this vast programme the policy has been that of testing and proving all smaller entities before incorporating them in larger ones. This is a long road starting with individual valves, pipes, electronic packages and pieces of structure, then testing hydraulic, fuel, electrical and other systems separately, then in co-ordination, and so on through a gradual process of agglomeration until very large entities are ready for test, such as the rocket engines, complete rocket stages and finally, complete rocket vehicles.

For the rocket engines special firing 'pits' were built in the Susana Mountains of California for the H-1 engines and in the Mojave Desert near Los Angeles for the F-1 engines. During the development stages these and other engines were put through hundreds of tests. The complete rocket stages are mounted in stands–that for the S-IC stage is 407 ft high–at the Mississippi Test Facility for general tests, including firing of all engines, before shipment to Kennedy Space Center. And, naturally, great care is taken in assembly of the launch vehicle stages, in launch pad checks and in count-down procedures.

This policy has undoubtedly paid off and, at the time of writing, about fifteen Saturn I and IB vehicles and six Saturn V vehicles have been launched successfully. This is no mean feat when one considers that only a few years ago four successful launches out of five was considered good for much smaller launchers.

Similar tactics were adopted for the various spacecraft modules and at the Kennedy Space Center special chambers have been built at the Manned Spacecraft Center's Space Environment Simulation Laboratory, in which tests can be made on equipment in the thermal and vacuum conditions of space. For example, before the first manned Command Module–Apollo 7–was launched in October 1968, an Apollo spacecraft consisting of both Command and Service Modules and a three-man crew were subjected to an eight-day test in simulated space conditions. Apart from checking all the various systems on board and demonstrating their adequacy, these tests were used to prove that the Earth recovery-system equipment and the ablative shield structure would stand up to the cold soak conditions of space. Such testing facilities are essential with this kind of programme, for in the coldness and vacuum of space, materials, surfaces, lubricants and so on often behave quite differently from the way they do on Earth. The only way to have confidence that something will work for days in space is to work it for days in a simulated space environment.

This kind of testing–for all equipment–goes on continually as refinements and modifications are made during the development period, and it has led to a far higher level of safety and reliability than many considered possible even a few years ago. Consequently, the fire which occurred in the spacecraft in early 1967 came as a nasty shock.

At the time, with work going well, hopes were high and the first manned Apollo spacecraft was being readied for flight on a Saturn IB vehicle at Complex 34–discussed earlier. In a 'Plugs Out Test'–a complete launch rehearsal in which umbilicals are disconnected–three astronauts died in a Command Module fire on 27th January 1967. They were Virgil I. Grissom, the second American in space, Edward H. White, famous for his 'walk in space', and Roger B. Chaffee, a relative newcomer.

Quite suddenly the astronauts reported a fire in the spacecraft, but this spread so rapidly that, before the crew on the inside or the launch personnel on the outside could open the spacecraft hatch, the occupants were dead and the Command Module was devastated.

Although the spacecraft was later examined in great detail by engineers from NASA and contractors, there was insufficient evidence to nail the cause down firmly. However, the seat of the fire was believed to have been in the lower forward section of the left-hand equipment bay, below the Environmental Control Unit, and the probable initiator was an electric arc in a power cable in that area. In any case, a repetition had to be avoided, not only for the sake of the astronauts but also in order that the whole project might not fail. It would be tedious to spell out all the lines of action taken, but there were a number of different approaches.

The first line of approach was to make it easier and quicker for the astronauts to get out of the spacecraft while participating in tests or waiting for a launch. This led to a completely redesigned Command Module Hatch which, though pressure-tight and incapable of being opened accidentally, could be opened from inside in seven seconds and from outside in 10 seconds. Escape routes were also checked, added to and obstructions removed.

The second line was to minimise the effects of a fire by providing better on the spot fire-fighting equipment, fan-extractors to remove smoke from the 'White Room' on the spacecraft access arm, and to fit water sprays to cool the spacecraft and the launch-escape system on top, because the latter contained solid propellents which could be ignited by heat.

The third line of approach was a long one and it amounted to examining in detail all the equipment in the Command Module with a view to eliminating all possible sources of combustion as far as possible, and the replacement of combustible materials with inert materials where practicable. This led to replacement of plastics switches with metal ones, with substituting stainless steel for aluminium in high-pressure oxygen tubing, with providing protective covers over wiring bundles, and armour-plating for water-glycol liquid line solder joints, and providing a host of similar modifications, including flameproof storage containers for flammable materials such as tissues, notebooks, charts, cameras and so on.

The last line of approach was concerned with the atmosphere in the spacecraft. We have seen earlier how it was decided to use an atmosphere of pure oxygen at 5 lbf/in^2 as standard in American spacecraft and this had proved quite satisfactory, providing sufficient oxygen for the astronauts to exist comfortably, without highly stressing the spacecraft or pressure suits and without constituting a special fire risk. However, as the spacecraft were not designed to take external pressure, when tests were being carried out at ground level the internal pressure was raised to a little above atmospheric–16 lbf/in². While pure oxygen at this pressure does not cause fires, it does lead to any fire spreading much more rapidly than in ordinary air–where the oxygen is diluted with four-fifths nitrogen–giving what is known as a 'flash fire'.

In order to ascertain the risks that might be involved, mock-ups of a Command Module cabin were made and fires were deliberately started under various conditions. In particular, extensive flammability tests were conducted with varying percentages of oxygen and nitrogen at different pressures. In the end, it was decided not to create an unnecessary fire risk at pre-launch testing, and a mixture of 60% oxygen and 40% nitrogen is now used instead of 100% oxygen at 16 lbf/in². This 'enriched air' is supplied by ground equipment and involved no changes in the spacecraft hardware. The crew suit loops, however, continue to carry 100% oxygen. After launch, as the rocket vehicle rises, the pressure in the spacecraft is gradually reduced to 5 lbf/in², and the 'enriched air' is replaced with the 100% oxygen. This takes a few hours and the crew remain in their pressure suits until this state is reached.

The fire risk is much less in space, partly because of the lower oxygen pressure and partly because in an emergency the crew could don their space suits and evacuate the cabin of oxygen by opening the hatch to the vacuum of space, which would douse most fires.

At the time, this accident came as a great shock and public reaction added greatly to the impact. As Dr von Braun pointed out later, it was a strange situation–after an uninterrupted string of successful flights in Mercury and Gemini spacecraft logging 1900 hours and 35 million miles in space without mishap, three lives were lost on the ground. If, he said, this had been an accident with an experimental aircraft where accidents occur rather frequently, it would have been accepted as the tragic outcome of some routine testing of a new piece of equipment.

One of the natural tendencies was for an 'over-reaction'– for everyone to re-examine everything and order modifications for anything that anyone might think remotely suspect in an endeavour to 'plug all gaps'. This, of course, can never be done and one quickly moves into an area of rapidly reducing returns for expended effort. The astronauts themselves, being qualified engineers and test pilots, understood right from the start that there existed the possibility of a sudden accident and that this risk will always be there, and they have done much to help put the matter in perspective and avoid over-correction.

All the same, the modifications put in hand to the Command Module were eating away at margins. In 1963 the Command Module had a drawing-board weight of 9500 lb. By 1967 it had reached 11,000 lb, and by the time eventually of the first manned Apollo flight–Apollo 7–this figure had grown to 12,600 lb. This heavier weight made it necessary to modify the drogue and main parachutes in the Apollo Earth-landing System. The drogue chutes diameters were increased from 13·7 to 16·5 ft, and two-stage reefing in the 83·3-ft diameter main parachutes was introduced to provide three phases of inflation–reefed, partially reefed and fully opened–to lessen opening shocks.

The implications of the Apollo spacecraft fire applied equally well to the Lunar Module as to the Command Module so that there was much heart-searching and modification in that area too, with great efforts simultaneously being made to reduce weight.

Indeed, a report to Congress at the end of 1967 said that spacecraft weight was the most critical problem in the programme at that time. The nominal weight for the complete Apollo spacecraft was early fixed at 95,000 lb–not including the Launch Escape Tower of 8,000 lb and the Spacecraft-LM Adapter of 4000 lb, both of which were jettisoned at early stages and could be considered as part of the Saturn vehicle so far as weight was concerned. This overall weight was shared out and by 1967 11,000 lb were allocated to the Command Module, 52,000 lb to the Service Module (10,200 lb dry), and 32,000 lb to the Lunar Module. The engineers concerned were having great difficulty in keeping down to these figures and when it became necessary to incorporate modifications as a result of the fire, the situation would have become even worse but for the fact that improved performances in the rocket engines allowed the total spacecraft weight to be raised somewhat.

For the lunar-landing mission–as distinct from earlier missions where weight stipulations were not so critical–the weight allowance for the Command Module had been increased from 11,000 to 12,600 lb, and at the end of 1967 engineers were already estimating within 78 lb of this weight limit and this was not considered a big enough margin for changes bound to arise in the meantime. The Service Module weight was also close to its limit, and that of the Lunar Module was already 42 lb overweight.

To counteract the inevitable demands for modifications which would increase weights yet more, engineers were examining every item of equipment for its weight-saving possibilities–changes of materials, changes of manufacturing method, changes of size and so on. At the same time task teams were also appointed to look at the possibilities of reducing the amount of cabin insulation, trading off stowage locations among the modules, and flying lower orbits around the Moon, all of which could save weight.

At this time, the fire seemed a more serious setback than it later turned out to be. The first manned flight of an Apollo spacecraft was delayed from about February 1967 to October 1968, and the estimated cost of the fire was finally reckoned at about $66 million or £27½ million–£15 million for design changes and equipment modifications in the spacecraft, £6 million for re-scheduling of deliveries of the Command and Lunar Modules, £2 million for new materials and flammability testing, £2½ million for development and testing of new fireproof flight suits, and £2 million for modifications to the launch facilities.

However, other work went ahead and, through rescheduling, two Saturn V vehicles with 'boiler-plate' Apollo spacecraft aboard were fired off before the first manned Apollo spacecraft was launched on a Saturn IB. The success of each of them matched up with the historic flight of Apollo 8 round the Moon, and the overall programme was then very much back to schedule.

The first firing of the Saturn V rocket vehicle took place on 9th November 1967, with a virtually flawless performance which raised the spirits tremendously of all those engaged in this project. For all those persons who had worked on the Saturn V and on the launch facilities this was the moment of truth. As the seconds ticked by and the great rocket soared into the sky, tension was released and tears streamed down the cheeks of grown men. The first stage fired for 2½ minutes, lifting the vehicle to 38½ miles height and a speed of 6100 mile/hour. The first stage was then jettisoned and the second stage pushed on to about 117 miles height and a speed of 15,300 mile/hour after 6·1 minutes firing time. After a burn of 2 minutes 12 seconds the third stage and the spacecraft went into Earth orbit and, after two orbits the third stage was re-ignited for about 5 minutes boosting the spacecraft to a speed of 23,370 mile/hour. The third stage was then jettisoned and the Service Module rocket engine was fired briefly, sending the Command Module to a distance of 11,286 miles. 8 hours 40 minutes after lift-off, after some additional rocket thrust to simulate a return from the Moon and the ditching of the Service Module, the Command Module hit the Earth's atmosphere at about 25,000 mile/hour and began its descent, landing in the Pacific only 10 miles from its predicted position.

For the first flight of the world's largest rocket vehicle to date, this was truly impressive. But what of the Launch Pad–how well did this stand up to the pounding it took from the Saturn V's blast envelope? On the whole, very well indeed. A broken joint on a pipe in the tower started a small fire, and the engine-servicing platform, which had been removed from under the blast opening to what was considered a safe distance, had been blasted a farther distance and was virtually a write-off. Of the parts we have discussed the only ones to suffer damage were the Tail Service Masts. Although the 'clam shell'-type hoods had been well tested in ordinary circumstances, when it came to the day, launch vibration or some other environmental factor caused the locking mechanisms to malfunction. The hoods on closing consequently rebounded, the engines' exhaust blast entered and completely destroyed the ground-carrier plates. As the next Saturn V launch was only a matter of weeks away, some quick action was needed here to prevent a reoccurrence. In fact, the hoods together with all their closing mechanism and hydraulics were abolished; in their place went the simplest imaginable protection. A lattice tower was erected on top of each Tail Service Mast at the top of which was mounted a hood with a vizor–very much like a knight's armoured helmet. The vizors are now raised and latched in position. When the Tail Service Masts operate, they swing upwards and when they have entered their respective hoods, the latches are tripped and the vizors slide down over their fronts so that all is covered. The only complication was that these small towers had to be made to hinge back and downwards when not in use, otherwise the Mobile Service Structure would foul them when in position on the Pad.

The Saturn V vehicle itself behaved very well indeed. The Apollo spacecraft and the third S-IVB stage had, of course, been tested earlier on Saturn IB flights, but this was the first flight for the lower two stages and, naturally, for the whole as a single unit. Accordingly, the engineers positioned measuring instruments all over the place; indeed, 2894 measurements were taken with this Apollo 4 flight, 972 in the first stage, 976 in the second, 609 in the third and 337 in the instrument unit, and these were sent back to waiting engineers by twenty-two telemetry systems. On board were two motion-picture cameras, these being mounted in the second stage to record the dual plane separations of the first stage and the interstage from the second stage. When their job was finished these cameras were ejected and descended by paraballoon for a splashdown in the Atlantic about 470 miles out, where they were recovered by teams homing on each camera's radio beacon.

Among all the measurements taken on the first S-IC stage there were eighty concerned with vibration, and engineers watched these especially. During the design stages, a stability model had been created to check on what has come to be called the 'Pogo loop'. Everything material has a natural frequency of vibration, which results from disturbing it. The disturbing force with a bell is the hammer or clapper which hits it, and its natural frequency is the number of cycles/second we hear as the sound coming off. In general, the larger the object or structure, the lower the natural frequency, and investigations showed that in the first stage of the Saturn V vehicle the natural frequencies of both the structure itself and the liquid oxygen suction duct would be in the region of 5 cycles/second. If now the engines also develop a similar frequency all in phase, then these make the structure vibrate, which through the liquid oxygen makes the suction ducts vibrate, which in turn inject the vibration back into the engines to complete the circuit or loop–and as the form of structural vibration is that which tends to make the rocket vehicle grow alternately longer then shorter, it has come to be known as the 'Pogo effect' after the Pogo stick. Vibrations like this which are in tune are said to be in resonance, going round the loop, continually reinforcing themselves so that the amplitudes grow bigger and bigger and, if they continue long enough, something quite catastrophic could happen–such as when a singer hits a note and breaks wine-glasses with the sound.

Against this tendency there is also a certain amount of damping in any structure which tends to reduce the amplitudes of vibration. In this first Saturn V flight, the engineers' predictions that there should not be any noticeable Pogo effect were confirmed and all was well.

Twenty-one weeks later the story was different. The measured oscillations were three times larger on this second Saturn V launch; the thrust oscillations of the F-1 engines were in phase during the latter part of the flight; there was structural feedback of the thrust vibrations into the propellent feed system; and quite significant longitudinal oscillations occurred at about 5·3 cycles/second. No harm was done, because the vibrations occurred late in the flight of the first stage, but it was an alarming situation.

A search was immediately made for a solution which would

give stability margins acceptable for a man-rated vehicle, and this had become important because it had just been decided that the third Saturn V vehicle would be manned–indeed, it was the historic Apollo 8 which sent Borman, Lovell and Anders round the Moon. This vehicle was already assembled in the Vehicle Assembly Building and was in the

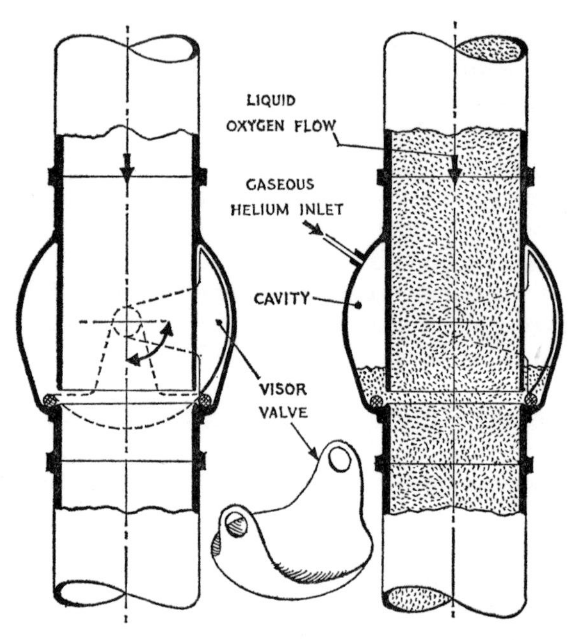

Fig. 43. The Prevalve in the Liquid Oxygen Ducts with the open and closed positions shown to the left. The right view shows the valve used as a gas accumulator to dampen out vibrations.

process of checking out. Consequently, the following stipulations were made. Implementation of a solution should not delay the launch schedule; the solution had to be analytically and experimentally demonstrable before any hardware was fitted; the effect on engine performance had to be negligible; and modifications to existing hardware had to be minimised.

Of course, NASA engineers at the Marshall Space Flight Center and Boeing engineers were familiar with this kind of problem in other rockets and they knew the kind of basic solutions available. Because of the above requirements, only two were deemed feasible and, after a lengthy analysis, one of these had to be rejected as, in curing instability at one frequency it might well introduce it at another. The most promising solution was that of adding a gas accumulator on each liquid oxygen suction line; this would introduce an element of 'springiness' into each line which would decouple the natural frequencies or break the loop.

This solution looked especially promising because of the pre-valves fitted into these ducts. These have no flight function and were included just in case there was a failure of a main oxygen valve during ground operations. What made these valves so interesting was that, as seen in Fig. 43, each had an internal annular volume surrounding the liquid oxygen flow path. This cavity housed the valve-closure element in the open position, and it was thought that filled with gas this cavity might serve as a spring-like pneumatic accumulator.

Tests showed that this indeed worked and all that was necessary to be carried out in terms of hardware was to connect a supply of helium–already available in the stage–to these pre-valves and arrange for the cavities to be filled before lift-off.

On the second Saturn V flight, designated Apollo 6, there was also a premature shut-down of the J-2 engine in the third stage. Analysis of data telemetered back indicated leaks in the igniter propellent lines of two J-2 engines–that of the third stage, and one in the second stage. Extensive ground testing led to redesign and 'beefing up' of these lines. They were tested in the Apollo 7 mission on the Saturn IB launch vehicle and passed as satisfactory.

These, then, are just some of the 'behind the scenes' problems and setbacks which have been solved and overcome to make man's first lunar flights so successful.

16

Countdowns, Guidance and Communications

SPACE travel has been a favourite topic of science-fiction writers from the time of Jules Verne onwards and some writers a whole generation ago made, in effect, predictions which match up quite recognisably with events, missions and hardware today. There is, however, an area which was quite ignored because it was under-estimated–indeed, without experience, it may not even have been recognised. This is the area of *control*. The amount of information transmitted and processed to make manned space missions possible is enough to make anyone not familiar with the scene boggle. Without the computer and modern telecommunications space travel would still be just a vision of the future.

We have already briefly seen that extensive computer studies were used to help decide which mode should be used to go to the Moon. Computers have also been relied upon for making very extensive and detailed calculations in the design stages of the rocket vehicles and much other equipment; and they have been important parts of many of the simulators built both for development and training purposes. While in theory many functions carried out by computers could be coped with by manual control, or by men using slide rules and simple desk calculators, in many cases the time taken would be far too long to be practicable, and this comes to a head when a Saturn V vehicle lifts off to begin a mission. Very many things in the rocket vehicle and spacecraft have to be monitored, decisions have to be taken, and corrections have to be calculated and applied in *real time*. By this we mean that these operations must be carried out very nearly instantaneously for immediate application, as distinct from, say, design calculations which can be coped with 'off line', as they say, at any convenient time.

The whole control complex comes in, of course, much earlier than this, even at the launch site. Next door to the Vehicle Assembly Building is the Launch Control Center, a four-storey building containing a number of control and firing rooms, each filled with rows of desks with displays and controls. This vast complex of equipment is not, of course, used just for a few minutes over the actual time of vehicle firing. It comes into use as soon as a Saturn V vehicle starts to grow in the Vehicle Assembly Building and is used–is, indeed, the very centre–for all the check-out procedures, rehearsals, and countdowns to lift-off, as well as for control of the rocket vehicle during the injection into Earth orbit stages. So much, in fact, goes on that it is difficult to give even an outline in a reasonable space.

To begin with, as the Saturn V vehicle grows on its Mobile Launcher in the Vehicle Assembly Building, various subsystems and systems have to be checked out. This means, in general terms, that generated inputs cause outputs, which have to be measured and compared with predictions and recorded results. As the build-up of a huge vehicle like the Saturn V takes place over a period of many months, a great many checks and calibrations are necessarily repeated from time to time, and this situation is ideal for computer control, the computers being both in the Launch Control Center and in the base of the Mobile Launcher.

During tests, stimuli are transmitted from a computer program to the appropriate stage or vehicle test point by input–output registers and data channels. Responses are returned by similar routes and, after processing, they can be displayed, stored on tape or printed out as required.

There is, of course, a fair amount of time for the ordinary check-out tests on the launch vehicle, because these are carried out at the pace of building the vehicle, but this is not so during the countdown to lift-off. The countdown is still essentially a testing program in which a check-list is used to carry out tests and operations in a required sequence. During the last few hours before lift-off, the 'launch-pad computer', as it is unofficially called, 'converses' continually with a twin computer in the Launch Control Center over a data link. As the computer at the pad goes about its business–commanding the exercise of valves, engine parts, relays in the vehicle and so on, and measuring their performances–it reports any problems arising to its electronic companion 3 miles away; this computer, with a different program, searches its memory for the corrective action and advises the pad computer what to do. Naturally, everything is under the surveillance of the control team engineers, who can break in at any time and take over control.

The situation has got to the point where complete vehicle launches can be simulated by computers, so that engineers can suggest better approaches and sequences of events, and can try them out without putting the actual launch of a vehicle to risk. And NASA actually uses a number of computers to compile programs for other computers to use . . .!

When the Saturn V has actually lifted-off, it has to put the spacecraft and the third stage into the Earth orbit planned for them. A small computer in the Instrument Unit of the rocket vehicle virtually controls this. It contains the program which, by gimballing the engines, steers the vehicle on the correct course, separates the stages and fires the engines of the second and third stages at the right times. This is not just a straightforward obeying of instructions, but a very precise guidance resulting from continual feedbacks. The attitude of the rocket vehicle is measured continually by reference to a 'stable platform', which is gyro-stabilised to a space reference. This platform contains accelerometers, which continually measure accelerations along the X, Y and Z axes. The information from these sources is compared with the programmed data so that corrections can be applied when necessary. Wind drift is noted but corrections are not put into effect until the vehicle is out of the lower atmosphere.

Minor corrections are being made all the time, but big corrections might need to be coped with at times. For example, one of the F-1 engines might fail during flight, or be shut-down to save an accident. Although there would then be more fuel left for the other four engines to use, so that they could burn longer, there would be a fifth drop in thrust and, depending upon precisely when the shut-down occurred, the action to be taken would be quickly computed and put into effect.

The immediate aim at the time of launch is to get the vehicle and the astronauts safely into their planned Earth orbit; from this position of comparative safety the situation can be assessed ready for the next stage of the mission. If very serious corrective action had to be taken during the launch sequence, the third stage could well have used up too much of its fuel, making the trip to the Moon impossible. Some lesser objective would then be selected or the mission would be cut short and the astronauts returned to Earth.

Assuming that the spacecraft is safely in orbit, it will then be boosted into a trans-lunar trajectory and the third stage will later be jettisoned. The spacecraft is now controlled by a computer on board linked by radio to a computer on Earth using as input data originating from Earth or put in locally by the astronauts.

Although the original Earth orbit of the spacecraft would be known from radar and other observations with a fairly high degree of accuracy, no measurements of position can be 100% accurate, nor can measurements of thrust and acceleration. These parameters can only be known within the limits of accuracy of the instruments and equipment used.

If we know the *exact* position and speed of a spacecraft at a particular time, we can predict its position and speed at any required future time, since motion in gravitational fields obeys simple laws. However, when the position and speed are known only within limits, as must always be the case, the limits continually increase as time goes by unless further measurements are taken to redefine positions more accurately. Really accurate determinations of velocity in a gravitational field can only be made by noting the position of the spacecraft over a period of time.

Position can be established by using radar from Earth, and with the big dishes of radio telescopes now available, such measurements are, in fact, accurate enough to be able to guide a spacecraft to the Moon and land it within a few miles of any given position. Using ordinary passive radar the received signals would be weak, though usable. These days, transponders are fitted to spacecraft as they are to aircraft. These are triggered by a radar signal and broadcast their own signal back, this being some thousands of times stronger than a passive echo.

Much more accurate navigation can be accomplished by using optical sightings and the astronauts in the Apollo spacecraft have a special instrument incorporating a telescope and sextant for this purpose. With this they can measure very accurately the angles between any of twenty-two prominent stars and specific points on the Earth or the Moon. On Earth, surveying and navigation are essentially two-dimensional, and two angular measurements from known landmarks are sufficient to fix a position. In the three dimensions of space, three such measurements have to be taken as a minimum. When the sextant is aligned, the crew member presses a button which injects the readings into the Command Module Navigation Computer.

It is important to remember that, just as at the launch pad the local computer and that in the Launch Control Center were linked by a data line, so the Apollo spacecraft computer is continuously linked to an Earth-bound computer. Even if the astronauts did not put in optical fixes on stars, their computer would still receive navigational information derived from Earth radar. Optical measurements, however, become increasingly important as the spacecraft nears the Moon and the actual orbit round the Moon is determined through a series of sightings of lunar landmarks. The accurate determination of lunar orbits has assumed greater importance than was originally expected, because measurements made by the crew of Apollo 8 suggested that the Moon is not homogeneous, but is rather like a 'currant bun' with large patchy deposits of heavy metals in places which distort the orbit from the simple theoretical shape.

Small changes in spacecraft velocity can be made by using the attitude thrusters, but for large changes the Service Module engine is used. It is, of course, important that the engine thrust should be in the right direction. Not only must the spacecraft be correctly oriented, but the line of thrust must be through the spacecraft's centre of gravity, otherwise the craft would veer off course. To ensure this, the engine is made steerable and it is controlled by another stable platform and group of accelerometers, which can measure any tendency to move off course and put in a correction.

What comes as rather a surprise to most people is that the spacecraft out in space, orbiting the Moon or even on the lunar surface, is just as much in contact with Mission Control as when it was in Earth orbit or even on the Launch Pad. Apart from voice and television contact–the communication aspects most apparent to the ordinary man–there is a continual two-way communication between the controllers and the spacecraft. In the Apollo Modules hundreds of measurements are taken continuously–such as oxygen pressure, cabin temperature, crew members' heart-beats and breathing rates, states of equipment, fuels, electronics and so on–and these are reproduced on the displays of the relevant control consoles on Earth. A doctor can keep an eye on the astronauts' general fitness at any time, just as an engineer can see whether any part of the spacecraft is not functioning correctly. Indeed, warning lights and buzzers ensure that attention is called to such things when they reach prescribed limits. This kind of information is referred to as 'down data'. 'Up data' is not so prolific and as well as information it contains command signals to put various operations into effect at the appropriate times.

The communications systems used are very sophisticated and are beyond our scope here. However, to a first approximation, the process is as follows. All the instruments taking measurements on the spacecraft are scanned–that is, looked at one after the other, until the process is completed, when the examination starts all over again. The measurements from each instrument are translated into a code analogous to the dots and dashes of the Morse Code, but effectively a series of pulses similar to the binary system used on computers. The number of 'digits' allocated to any instrument depends upon the information and accuracy of reporting required. A channel signalling merely that some equipment is on or off need be very few pulses wide; one communicating a very variable instrument reading to a high degree of accuracy, needs to be very many pulses wide. A synchronisation signal is, of course, given at every cycle so that the receivers and transmitters are in step and so that the Earth computers allocate the right pulses to the right channels and correctly process the information. 'Down data' is transmitted at the rate of around 12,800 binary digits/second and the 'housekeeping' sampling, as it is called, is repeated many times a second.

Signals and commands going up to the spacecraft are, by comparison, only occasional and unpredictable, so that each is preceded by its 'address', that is, a coded signal saying whether the message to follow is for the navigational computer or the Service Module engine ignition control, and so on, much as is the case with computer internal circuits.

It is all very well glibly talking about all this information passing between the spacecraft and Earth–and at times live television and voice–but a tremendous communications network has had to be built up by NASA to cope, because Earth satellites move right round the Earth and even for deep space probes the Earth is spinning on its axis, so that no single radio station can keep spacecraft in view all the time. This communications network is complex because it consists of a number of integrated networks used for different purposes–one for keeping track of the hundreds of Earth satellites and 'space debris' such as abandoned rocket stages and dummy spacecraft left in orbit; one for tracking deep space probes, such as those despatched to Venus and Mars; and one for manned space-flight missions.

So far as the latter is concerned, the network starts with a chain of radar stations spread out along the Atlantic Test Range, which are used primarily for monitoring the actual launch of the spacecraft into orbit. Control then moves from the Launch Control Center at Kennedy Space Center to Mission Control at Houston, Texas. Once in orbit, a chain of 30-ft diameter dish aerial antennas all round the world, including some in special ships, constantly receive data from the spacecraft and relay this to Mission Control.

When the Apollo spacecraft moves out into space on its way to the Moon, at about 10,000 miles out communications are switched to three large 85 ft diameter dishes situated roughly 120° apart round the Earth, so that as the world rotates at least one dish is always in radio view of the spacecraft. One is at Goldstone in the USA, one is near Madrid in Spain, and the other is at Canberra in Australia. Data received at these stations are initially processed and then passed into the communications network, passing by land line or communications satellite through switching centres to Mission Control at Houston. The biggest switching centre is the Goddard Space Flight Center, which receives all data originating from all spacecraft, satellites and probes, processes them in its computers and directs them to the appropriate control centre.

Similarly, information and commands from all control centres (including those for other NASA space activities) are routed via Goddard, where they can be ranked for priority and sent by appropriate routes. The reason for doing this is that it would be prohibitive for NASA to lay down its own communications channels linking relay stations all round the world; instead it leases or buys time on commercial channels, and some of these channels can deal with information at much greater densities than others, which means that they can transmit messages faster. The ranking of data received at Goddard consequently allows urgent messages to go by the quickest routes, while more routine information travels by slower paths.

Knowing about the continual 'conversation' going on between the computers on Earth and those on the spacecraft, one can understand why there is a kind of apprehension at Mission Control when an Apollo spacecraft disappears behind the Moon. Not only is voice contact lost, but for 45 minutes in each orbit, the craft and the astronauts really are on their own with all links to Earth completely severed.

17

The Apollo Series of Missions

WE have now discussed the many and varied elements of the Apollo project–the spacecraft modules, the launch vehicles and the ground systems and support equipment–and the time has come to recount the missions which used this equipment. The object, of course, is clear; it is the landing of two men on the Moon and their safe return with their third crew member to Earth. However, it is obvious that with so much new equipment and such complicated new systems it would be unrealistic to put all the major elements together with a view to achieving this objective the very first time. The objective has to be approached in several steps. A progressive series of tests were, therefore, planned prior to the actual lunar mission. These were not only to test the hardware and systems, but also to exercise and acclimatise the astronaut crews so, according to the nature of the missions, some were manned and some were unmanned.

The first in-flight combination of the Apollo spacecraft and the Saturn V vehicle was the Apollo 4 mission, launched on 9th November 1967 from the Kennedy Space Center. The profile for this mission has already been recounted in Chapter 15 and will not be repeated in detail here. As well as being the first firing of the Saturn V vehicle, this was a major test for the Command Module. The spacecraft was hurled over 11,000 miles out into space and, on its return, the Service Module engine was relit to increase its acceleration to equal that which it would have attained if it had returned from the Moon. The Command Module entered the Earth's atmosphere at 25,000 miles/hour and landed successfully in the Pacific Ocean. Indeed, the whole of this mission from every aspect was a brilliant success.

As a result of programme re-scheduling, there were officially no Apollo 1, 2 or 3 missions preceding this, but two unmanned Apollo spacecraft were tested in sub-orbital flights on 26th February and 25th August 1966, using Saturn IB launch vehicles, and the first planned flight with a crew did not take place as a result of the disastrous Command Module fire, which occurred during a ground test at Cape Kennedy on 27th January 1967.

The Apollo 4 mission carried out into space a Lunar Module Test Article–a 'boiler-plate' Lunar Module–which was attached to the third stage of the Saturn V. It had no landing gear, and the propellent tanks were filled with water-glycol and with freon to simulate fuel and oxidiser. The ascent stage was just a ballasted aluminium structure. Measurements of vibration, acoustics and structural integrity were taken at thirty-six points and telemetered back to Earth.

The next was Apollo 5, launched on 22nd January 1968. This used a Saturn IB launch vehicle and was specially to test in space a complete, though unmanned, Lunar Module. The apex of the Lunar Module was covered with a simple aerodynamic shroud, which was jettisoned once the vehicle was in orbit. The LM-Adapter Panels were deployed as on a lunar mission, and the descent stage engine was fired three times satisfactorily. At the end of the last burn, the two stages of the Lunar Module were separated by igniting the ascent stage engine. All the information from these tests was telemetered through radio ground stations and showed the tests to be successful. The two parts of the Lunar Module were left in orbits which quickly decayed–that is, through air resistance, became lower and lower–the ascent stage re-entering some 4 days later and the descent stage a few weeks later. Both, of course, burnt up, not being able to withstand the heat of re-entry.

The Command and Lunar Modules used in these tests were early versions, the final models taking account of the changes which were clearly needed after the fire in 1967 which took the lives of the three astronauts on the launch pad. The Lunar Module in the Apollo 5 test was considerably overweight compared with the version which finally landed on the Moon.

The next in the series of tests was Apollo 6 launched on 4th April 1968 and injected into orbit this time by a Saturn

V–only the second firing of this huge rocket vehicle. As mentioned in Chapter 15 there were some malfunctions in the launch vehicle–in particular, there was a loss of power in one of the second stage J-2 engines, and the engine of the third stage cut-out prematurely and failed to restart. Failures of this kind are, so far as is possible, anticipated and an alternative mission was quickly selected. While some major objectives were thus not achieved, the tests with the Command and Service Modules were highly successful. The navigation and guidance system was tested in conjunction with the Service Module Propulsion Engine, which had its longest burn to date–7 minutes 25 seconds–and, as well as another full test of the Command Module's heat shield, extensive tests were carried out on attitude control, using the spacecraft's thrusters.

The next test, and the first of the major ones, came with Apollo 7 launched on 11th October 1968 when for the first time a three-man crew occupied the Command Module. No Lunar Module was carried on this trip and, as only Command and Service Modules were being put into a low Earth orbit, a Saturn IB vehicle was suitable for the launch. The crew were Captain Walter Schirra, Major Donn Eisele and a civilian, Walter Cunningham. The Saturn IB put them into a circular orbit about 150 miles altitude and the spacecraft separated from the second stage of the booster vehicle nearly 3 hours after lift-off. A target circle had been painted on this rocket stage, which itself was in a similar orbit, and the crew practised docking manoeuvres by closing in with their spacecraft to within a few feet of this circle. The flight lasted 11 days, during which the Command and Service Modules were put fully through their paces. They performed extremely well, giving great confidence in their design. The crew, of course, also put up a wonderful show, demonstrating that three men could live and operate in the Apollo spacecraft for the period of time needed for the Lunar Landing mission.

The scene was thus set for the first of the deeper space tests, which came with Apollo 8–the flight which made history when three American astronauts circumnavigated the Moon over Christmas 1968. This flight was not, in fact, one of those originally planned–neither was the Saturn V vehicle suitable for a manned mission–that is, its acceleration, vibration and other characteristics would be 'rough' on the crew over the launch period. However, after the Apollo 6 flight in April, the decision was made that this would be the first Saturn V vehicle to put men into space. So the stages were dismantled, and the second stage was returned to the Marshall Space Flight Center Mississippi Test Facility for various modifications to be made. At the same time modifications were being made at Kennedy Space Center on the J-2 engine of the third stage, and on the liquid oxygen pre-valves in the first stage to overcome the troubles encountered on the previous Saturn V launch.

The Saturn V vehicle was fully reassembled in August and the spacecraft was ready for fitting. However, although Lunar Module 3 was to have been incorporated, this was delayed as the result of weight and other troubles, so that another Lunar Module Test Article was incorporated instead, the weight of which could be adjusted to suit the mission.

The object of the mission was to complete ten orbits of the Moon without landing and then to return to Earth. Originally it had been expected to fly the Lunar Module manned in Earth orbit at this point in the series of tests and missions, but events suggested that this mission would be later than planned. However, the Apollo missions were arranged flexibly so that one could 'leap-frog' over the other if need be and at this time it seemed expedient for technical and political reasons to advance the first flight to the Moon.

The crew of Apollo 8 were Colonel Frank Borman (Commander), Captain James A. Lovell (Command Module Pilot) and Major William A. Anders (Lunar Module Pilot). These three men were to make history by being the first humans to leave the effective gravitational field of the Earth and circle the Moon, which they did in a most dramatic and successful way over Christmas 1968.

The Apollo 8 mission will be remembered not only for its demonstration of the dramatic advance in space technology, but also for the incredible perfection which was achieved throughout the whole mission by both the equipment and the men–in space and on the Earth. Ironically, some human problems were encountered, in that Colonel Borman in particular suffered during the outward flight, and to a lesser extent on the return flight, some form of sickness which appeared to be related to taking Seconal sleeping pills (to which he was not accustomed). This led to a feeling of nausea and, as there had been a slight epidemic of influenza in the Kennedy Space Center Area, there was some concern to start with that the astronauts might be suffering from this illness – following the bout of influenza suffered by all the Apollo 7 crew members during their mission. Fortunately this proved not to be the case, and all the crew were in good physical form to complete the mission successfully.

The official objectives of the mission were, of course, rather more detailed than just that of making ten orbits round the Moon. They were to demonstrate the performance of a number of mission activities, including translunar injection, Command/Service Module navigation, communications and mid-course corrections, and to obtain an assessment of the spacecraft consumables and passive thermal control. In addition, detailed test objectives were designed to 'wring out' thoroughly systems and procedures that have a direct bearing on future lunar landings and other space operations in the vicinity of the Moon.

The mission, like all others, was carried out on a step-by-step 'commit point' basis, which means that decisions whether to continue the mission, return to Earth or change to an alternative mission are made before each major manoeuvre, depending upon the operational status of the spacecraft systems and crew. These alternate missions are numerous and complex, but in this case they amounted either to returning to Earth direct by way of an elongated elliptical path in space, which is effectively still an Earth orbit, albeit with the farthest point on the other side of the Moon, or a simple lunar fly-by without going into lunar orbit.

In fact, the flight was a faultless demonstration of the Command and Service Modules, particularly the restart capability of the main Service Module Propulsion engine upon which the return journey to Earth depended so critically.

The crew carried on board a television camera which, on many occasions, gave live television pictures–released by Houston Mission Control to the world's television networks –of both the crew inside the spacecraft and views looking down on the Moon from a mere 70–75 miles altitude. The crew also carried, as usual, various cameras loaded with colour film and these yielded for the first time most dramatic pictures of the Earth viewed from the Moon, and photographs of the Moon's surface as seen by the astronauts in place of the hitherto remote pictures received from black and white cameras.

The Apollo 8 mission will be remembered too for bringing to the world's attention the tremendous achievement in the *management* task of assembling the equipment, of controlling all the ground support facilities of communications and tracking and recovery, and of achieving a lift-off within one-sixth of a second of that planned many months before. Such management expertise was necessary for the previous missions too, but it took an historic achievement of this nature to bring this into general recognition.

Several important things, of course, were learnt from the flight, which led to modifications to subsequent plans. For example, in talking to one of the authors Colonel Borman mentioned the fatigue which the crew felt when they reached the vicinity of the Moon, having decelerated into Lunar orbit by using their Service Module engine. While looking down upon the surface of the Moon they realised what an extremely tiring and difficult task it would be for subsequent crews who, at that point of the flight, had to climb through the tunnel into the Lunar Module, separate, descend to the Moon's surface and complete their task. As Frank Borman remarked to Jim Lovell, who commented at that time on their tiredness, 'It would certainly be some chore,' for the men who would make the final descent. This emphasised the need to conserve the mental and physical energy of the astronauts, who are not, as some people believe, supermen who never get tired. It is essential for even relatively short missions to the Moon, to ensure that the men who have the final task of descending to the surface are well rested, and this experience has led to revisions in the work schedules of the two crew members who make the actual landing.

This, of course, is all part of the learning process, and one of the reasons for running preliminary missions before the actual Lunar Landing flight. In other respects, the Apollo 8 mission was tremendously successful in demonstrating the fine integration of men and machines.

Precision was again the order of the day with Apollo 9, which splashed down in the Atlantic only 10 seconds late after its 10-day mission in space. The crew were Colonel James A. McDivitt, in command, with Colonel David R. Scott and Russell Schweickart, the civilian Lunar Module pilot. The complete three module Apollo spacecraft was blasted off from Cape Kennedy on 3rd March 1969 atop a Saturn V launch vehicle. This time the mission was confined to Earth orbit and the main exercise of the mission was to test the Lunar Module in space with men at the controls. The idea was to simulate the Lunar Landing mission as closely as possible though remaining about 120 miles above the Earth's surface all the time, and the plan was as follows.

First the Saturn V boosted the third rocket stage and the Apollo spacecraft into a near circular orbit at about 110 miles altitude. After a simulated insertion into a translunar trajectory, the LM-Adapter panels blew apart, exposing the Lunar Module and leaving the Command/Service Module free. Using its thrusters, the Command/Service Module turned through 180 degrees and then moved in towards the Lunar Module, using the Command Module's docking probe to guide the two craft together until they were joined by their docking hatches. During the remainder of the first day and the second day the three modules locked together were placed into various different orbits–reaching an altitude of about 300 miles–by burning the Service Propulsion Engine in simulated course correction manoeuvres.

On the third day, McDivitt and Schweickart, as Commander and Lunar Module Pilot, equalised oxygen pressure in the two spacecraft, opened the docking hatches and climbed through the tunnel for a 3-hour check of the Lunar Module systems and equipment, including a brief test firing of the descent section propulsion engine, before returning to the Command Module.

On the fourth day there was some extravehicular activity and television operations, while the crucial test came on the fifth day. McDivitt and Schweickart re-entered the Lunar Module and closed their docking hatch. Scott likewise closed the Command Module docking hatch, and the two spacecraft separated. The descent stage engine was then ignited to move the Lunar Module into an orbit which took it nearly 100 miles away from the orbiting Command Module. At the appropriate time the two Lunar Module stages were separated and the ascent stage engine then used to effect a rendezvous with the Command Module. After docking and transfer of the crew, the Lunar Module was jettisoned and from there on the procedure was as for other missions.

This was a very critical mission for, if the Lunar Module had failed in its various operations not only would it have seriously jeopardised the lives of its crew, but it would have brought the Apollo programme to a shuddering–though no doubt temporary–halt.

As it was, except for a slight curtailment of the extravehicular activity as a result of Schweickart suffering a short-lived indisposition the whole mission went very much according to plan. The separation and functioning of the ascent and descent stages of the Lunar Module were faultless and during the extravehicular activity, Schweickart actually stood on the steps of the ladder mounted on one leg of the Lunar Module, simulating as it were the first astronaut about to land on the Moon.

The scene was now properly set for the culmination of the whole project. So successful was the Apollo 9 mission that there was immediate speculation as to whether NASA would choose to make Apollo 10 the Lunar Landing mission. In fact, they decided not to 'push their luck' by advancing too quickly. It was decided that Apollo 10 should be a combination of the operations tested in the two previous missions, that is a complete test of the two Lunar Module stages in orbit round the Moon, carrying out all operations for the major mission other than the actual landing.

The final dress rehearsal commenced on Sunday 18th May 1969, when Thomas P. Stafford (Commander), John W. Young (Command Module Pilot) and Eugene A. Cernan (Lunar Module Pilot) blasted off within a few seconds of the scheduled time of 12.49 local time. Eleven minutes after lift-off the third stage of the Saturn V rocket vehicle and the spacecraft were in Earth orbit. Two orbits later the third stage fired again for 5 min 22 sec to put the spacecraft into the required translunar trajectory.

As with Apollo 9, the astronauts detached the Command and Service Modules by releasing the LM Adapter panels and exposing the Lunar Module. Their manoeuvre of turning around and docking with the Lunar Module was, in fact, transmitted live by television not only to Houston but to most of the world's television networks. Although the Saturn third stage was jettisoned, the next day the crew reported sighting 'a twirling object'.

After checking, Mission Control reported that it could only be this third-stage rocket casing following on behind. What was surprising–if this was indeed the case–was that this stage was then some 3000 miles behind the spacecraft! This was yet another example of the extraordinary clarity of vision all astronauts have reported in the space environment.

On arrival at the Moon–a few minutes later than scheduled –the spacecraft was put into the correct lunar orbit. The next day, Stafford and Cernan climbed through the tunnel connecting the modules, and made ready to separate 'Snoopy', as the Lunar Module this time was code-named. However, when both the Command and Lunar Module hatches were replaced it proved impossible to evacuate the air from the tunnel space between them–and it was necessary to do this to check that the hatches were, in fact, air-tight before module separation. A sticky valve proved to be the trouble, and by running through the routine again all was eventually well. After separation and a run through the check-out procedure, Mission Control gave Snoopy permission to 'go down' and, by firing the descent stage engine for 28 sec Snoopy was put into an orbit with a perigee only 9 miles above the area selected for the first lunar landing in the Sea of Tranquillity.

At this crucial point–the closest men had been to the Moon's surface–both still and ciné cameras jammed. However, it was later found that a fair amount of film had been correctly exposed during this manoeuvre.

At the appropriate time, as planned, the descent stage was jettisoned and the ascent engine ignited, simulating the eventual take-off from the lunar surface. As the two stages separated, for a few brief moments the ascent stage developed a wild gyration through a switch being left in the wrong position before being brought under control. However, all went well and after a few hours adjusting orbits, the Lunar Module redocked with the orbiting Command Module, code-named 'Charlie Brown'. The two astronauts transferred back to the Command Module and the Lunar Module was jettisoned by putting it into a solar orbit. However, during later lunar orbits Stafford reported sighting this Lunar Module 'uncomfortably close'. Mission Control assured the astronauts that it would soon be hundreds of miles away and 'not to worry'.

The Command Module then continued to complete thirty-one lunar orbits before firing the Service Propulsion System engine to bring them back to Earth. Apart from photographic work, the crew obtained a large number of optical sightings to establish the spacecraft's orbit as accurately as possible. The reason for this was that local variations in the lunar gravitational field–believed to be due to 'mascons' or massive concentrations of heavy material below the Moon's surface in some places–caused Apollo 8 to be considerably out from its predicted position on a number of occasions. For instance, after two orbits the Apollo 8 spacecraft was found to be 6 miles out from its predicted position and 5500 ft out in height. Subsequent analysis of the spacecraft's path allowed more accurate predictions for Apollo 10, but the data collected on this mission was essential to get really good predictions for the landing of Apollo 11.

One of the remarkable aspects of this mission was the television coverage. The crew put a lot of effort into these broadcast periods and the technical achievement was little short of fantastic. Both black and white, and colour pictures were received as clear on millions of homes' screens as local broadcasts, though they were coming through a quarter of a million miles of space.

Apart from items already mentioned, a number of minor things went wrong–trouble with the fuel cells, chlorination of the water they produced, loose fibrous-insulating material floating weightless in the cabin atmosphere and so on–but the crew managed to cope successfully with all these faults and once again demonstrated how the versatility of man in space can work towards mission success.

The Apollo 10 Command Module eventually splashed down only 25 sec late in the Pacific on 26th May after a mission lasting 192 hours 3 min. So well had the mission gone that there now seemed no doubt that before July was out Man would have set foot on an alien world.

And so it proved to be, for at 03.56 British Summer Time (BST) on 21st July 1969 Neil A. Armstrong, the Apollo 11 Commander, placed a booted foot on the surface of the Moon, and so became the first human being to visit an extra-terrestrial body.

The Apollo 11 mission, building upon the experience of its predecessors, went extraordinarily well, beginning with a classical lift-off at 14.32 BST (8.32 CDT) on 16th July. Technically this was no different from any previous Saturn V launch, but the knowledge that it was to take the first men to the Moon attracted an unprecedented crowd to Cape Kennedy, while the lift-off was watched on television by hundreds of millions all over the world. The injection into Earth orbit followed closely the pattern of the Apollo 10 profile, as did the translunar trajectory burn and the docking of the Command Module with the Lunar Module.

Upon reaching the Moon on 19th July, the complete three-module spacecraft–which engineers had managed to cut down to a mass of 96,300 lb–was put into an elliptical orbit with its lowest and highest points being 60 and 170 miles above the lunar surface. About 4½ hours later this orbit was adjusted to a more circular one at heights varying between 54 and 66 miles. Twnety-one hours after entering lunar orbit, Armstrong and Edwin E. Aldrin checked out the

Lunar Module prior to undocking. On this mission the code name for the Lunar Module was 'Eagle' and that for the Command Module was 'Columbia'–more serious names than those used hitherto, Eagle being the animal representing America, while Columbia is the poetical title for America, derived from Columbus.

At the appointed time the two craft separated, Lt-Colonel Michael Collins, the Command Module Pilot, staying in orbit with Columbia, while Mr Armstrong and Colonel Aldrin put Eagle into an elliptical orbit which extended to within 10 miles of the Moon's surface about 260 miles up range from the planned touch-down point. All being well at this time, Eagle then went into a 'three-phase powered descent initiation', the braking manoeuvre which was scheduled to reduce the craft's velocity to zero at a height of around 7000 ft over the landing site. With 7 min to go and operations being checked at Houston Control, the crew received the signal, 'Continue your powered descent.' The landing was to have been virtually automatic until the actual touch-down stage, but the crew demands for data during the diving and landing phases became so heavy that the onboard computer approached saturation. This brought on a series of warning alarms, and Armstrong had to switch from automatic to semi-manual–with Aldrin calling out the velocity figures and Houston Control over-riding the saturated warning signals of the computer by confirming to the crew that the vehicle was still in a 'go' condition.

The Landing Radar apparently was not able to discriminate the surface contours sufficiently well to recognise that Eagle was headed to land in a shallow, but boulder-strewn, crater. Armstrong, seeing this through the Lunar Module windows, checked the rate of descent suddenly with a 32-sec full power burn of the descent stage engine, while the vehicle moved farther on to make a perfect landing just down-range from West Crater towards the edge of the Sea of Tranquillity. This was about 4 miles down-range of the predicted landing position as a result, it is thought, of a series of small cumulative errors in the various flight phases.

At 21.18 BST Neil Armstrong sent the message 'Houston,

Tranquillity Base here. The Eagle has landed.' And while Armstrong and Aldrin rested and prepared for their lunar exploration, millions on Earth waited by their television sets to see the first man step on to the Moon's surface. This–something quite unforeseen by science-fiction writers–was made possible through the inclusion of a $7\frac{1}{4}$-lb portable television camera in the Lunar Module's payload, a camera which was twice deleted during weight reduction programmes, but which luckily always managed somehow to get back again. This camera was mounted on the top of the package of scientific equipment stowed in the descent stage. After being 'talked down' through the hatch by Aldrin–the clearances of the combined man and Portable Life Support System being small–Armstrong pulled a wire which allowed the package to swing outwards and the camera to start shooting. Thus history was made while the whole world watched the Apollo 11 Commander slip from the ladder to the footpad and then step out on to the naked surface. This was at 03.56 BST, 109 hours 24 min into the mission.

Armstrong later moved the camera to a vantage point some yards away from the Lunar Module–Houston Control giving directions for aligning it–and in all this camera relayed over 5 hours of television pictures to Earth, including the final re-entry of the astronauts into the Lunar Module. This camera itself was a remarkable engineering achievement. To fit the weight and size requirements and the band-width of the data channel available, it could only televise at 10 frames/sec compared with the 25 frames/sec (Britain) or 30 frames/sec (USA) of commercial television. Storage circuits on Earth filled in the gaps to give continuous pictures, but when the astronauts moved quickly the lingering afterglow of previous images created a quite ghostly effect. The scan was also reduced to 325 lines so that the pictures were comparatively coarse.

So far as the astronauts were concerned, most of the surprises on the lunar surface were to their advantage. The bulky-looking space suits did not hinder movement as much as expected, and both men carried out a number of experiments to find the best method of locomotion–which turned out to be a long-paced loping walk. It seemed also that in the reduced lunar gravity the energy requirements for locomotion and work were significantly less than anticipated, so that body cooling was not problematical and each had plenty of consumables left when they had finished their excursion. They had spent a total of 2 hours 47 min on the Portable Life Support Systems.

After closing the hatch at 06.11 BST, the cabin was re-pressurised while the crew rested and checked systems for lift-off. At 18.54 BST, Ground Control gave clearance for lift-off, Aldrin replied, 'Roger, understand, we're No. 1 on the runway,' and they were away. All systems worked well and Eagle redocked with Columbia at 22.35 BST–a little bit rough and 5 min late, but safely. From there on procedure was similar to that for Apollo 10, the Command Module splashing down at dawn in the Pacific on 24th July, after a mission lasting 8 days 3 hours 18 min. This time, the space-craft turned over, apex down in the sea, and the crew had to inflate the special flotation bags to turn it over right way up. Armstrong, Aldrin and Collins were hustled into a special quarantine 'caravan' on the USS *Hornet*, greeted through glass by President Nixon and later flown–still in their caravan–to Houston.

The astronauts left on the Moon not only the Lunar Module descent stage and the ribbed footprints of their boots but also equipment to record lunar tremors, and a special cube-corner reflector for laser measurements of the Moon's distance at various points in its orbit. They brought back a solar wind experiment, about 60 lb of lunar rock and soil and, of course, photographs.

On top of this pure scientific effort were superimposed two other views. One was the emotional and symbolic view which goes with a 'first', especially one of this uniqueness–the fanfares, speeches and red carpets one naturally expects on such occasions; the other was the quiet acknowledgement, by those with the knowledge to understand, of the sheer technical achievement which this lunar landing represented and the high level of professionalism of everyone engaged in the operation.

Many more courageous men will make landings on the Moon, but with Apollo 11 the initial goal has been reached, and the conception and design of the spacecraft, launch vehicles, systems and support equipment have been vindicated beyond a shadow of doubt.

Project Apollo has not only broadened Man's horizon by opening up true space travel; it has also made us aware that with the right aims and support, Mankind can organise itself and its technology to achieve great things.

18
What Follows Apollo?

So men have now landed on the Moon and one of the most fascinating and tantalising objectives of man since the beginning of history has been accomplished. What next?

In the immediate future much thought has been given to what is called the Apollo Applications Program. The intention here has always been to use, to the greatest extent possible, existing hardware and to use the launch vehicles, spacecraft and ground launch equipment which have already been developed for the Apollo project. The United States now has a very large range of booster-rocket vehicles for putting up payloads of various sizes for a multitude of missions. While the step between the Saturn IB and Saturn V vehicles is rather large, intermediate projects could be coped with either by using the Saturn V first and second stages only, or by upgrading the Saturn IB by 'strapping on' additional rocket boosters, possibly of the solid fuel type.

Staff studies by the sub-committees of NASA prepared for the Committee of Sciences and Astronautics give the basic objectives of the Apollo Applications Program:

(*a*) long-duration space flight of men and associated systems;
(*b*) scientific investigations in Earth orbit;
(*c*) applications in Earth orbit;
(*d*) extended lunar exploration; and
(*e*) an approach to the development of a defined basis for potential future space programmes.

One of the obvious objectives has indeed been the creation of large laboratories in Earth orbit to examine habitability, bio-medical problems, and to make behavioural experiments. It has always been thought that one of the first major applications of a manned space station from a scientific point of view would be in solar astronomy, although astronomical observations of the Earth and the stars would also be performed from this vantage point outside the Earth's atmosphere.

Many people have considered that one of the most interesting applications of a large manned space station would be the extension of the present generation of unmanned satellites used for meteorological Earth resource and communication purposes. The present diversification of such spacecraft and the specialised equipment which is located in each follows to some extent the philosophy of 'not too many eggs in one basket'! In the event of a major failure, for example, of an attitude control or power system, an unmanned spacecraft or satellite may well be lost and another have to be sent up to take its place–but there would be some consolation in that it was providing only one specific service and not a whole range of services. However, with the coming of a manned space station, repair and maintenance becomes a far more practical proposition. With a service engineer in space, a module can be replaced, and items can be repaired and so on. This leads to the situation where one has to take a hard look at the trade-off between the complications and costs of having a man on board, but with the advantages of combining perhaps several different activities into a common spacecraft, and not having to write-off the craft as the result of a minor fault. Consequently, the proper relationship between manned and unmanned satellite operations has to be examined very carefully in planning the Apollo Applications Program.

Some of the long duration testing of equipment and men must clearly be related to the next step, which is the semi-permanent–followed by the permanent–habitation of the Moon in a scientific colony; the long term aim being missions to Mars and other parts of the solar system.

To gain some idea of what would be involved it is worth noting the various possible manned missions to Mars that might take place in the next few decades. Indeed, the time-scale for starting such operations could well be between 1980 and 1985. There are two main alternatives. The minimum total round trip flight time would be between 400 and 500 days, allowing a 10–20 days stop-over period at Mars. An alternative mission involving a much longer stop-over period on the planet might take up to 900 days or 2½ years to accomplish!

One of the most favoured missions to Mars includes a so-called 'swing-by' procedure whereby, on the way back from Mars, the trajectory is arranged to pass close to the planet Venus, allowing the spacecraft to be influenced by the Venusian gravity and swung into a completely new orbit before moving off along an hyperbolic trajectory back to Earth. This procedure–carried out when the three planets are in appropriate positions–would reduce the total rocket energy required for the return to Earth, would reduce the mission time to about 600 days and would, incidentally, give a 'free' close-up view of Venus.

It is not the purpose of this book to go into the various possibilities of manned trips round the solar system, but from these few remarks it can be seen that real interplanetary flights of the future are bound to involve periods of up to several years away from Earth. Although it may be possible to create some degree of artificial 'gravity' in a spacecraft–though this will probably be more a centrifugal force–some or perhaps most of these long periods will be in zero gravity environments. It seems that a completely different approach will have to be made towards providing the crew environment from that used for a 'mere journey to the Moon and back' which can be accomplished within a week! When talking to American astronauts one finds that they are already discussing and even planning for typically 400-day periods or more in a spacecraft. Leisure time in these circumstances can no longer be taken as an incidental, but it must be deliberately planned to ensure that a reasonable balance is maintained between concentrated, demanding work (in what is, after all, one of the most difficult environments one can imagine), and on the other hand relaxation periods which are seen to be essential on long missions. It will probably be necessary for a mini-gymnasium to be established in the spacecraft, or at least some facilities appropriate to it, as even a 2-week Earth orbital mission has shown deterioration in astronauts' muscles, especially in the legs.

Clearly, the Apollo Lunar Mission will have produced a tremendous amount of information without which it would not now be possible to contemplate so confidently these manned missions to Mars and elsewhere in the solar system. There will, however, be an intermediate step before the manned missions to the planets, and that is the establishment of the large Manned Orbiting Laboratory. Current consideration is being given by NASA to a module technique whereby before 1975 the beginnings of a station with perhaps twelve men on board will be achieved; further modules can then be added to build up the structure until it can accommodate up to perhaps a hundred men. The crew of such a station will be in orbit for lengthy periods of time–certainly weeks, possibly months or even a year–but there would be a regular shuttle service between the station and Earth, both for logistic refurbishing and the transfer of men.

The nature of this shuttle service is also being examined in depth, as a point is reached when it becomes more economic to replace the expendable launch vehicles, such as have been used for Project Apollo, by non-expendable, recoverable launch vehicles and spacecraft. Almost certainly the first truly recoverable and reuseable section of the complete configuration will be the actual spacecraft itself–after all, the Apollo Command Module is already recovered; and the extensive amount of work on 'lifting bodies' which has been carried out in the United States will obviously have a direct application here in allowing a hypersonic vehicle to descend from Earth orbit and land in a relatively conventional style on an airstrip, without the use of parachutes or sea recovery as at present used for Apollo.

The present series of Apollo Command Modules have returned from their missions in such fine shape that the makers are at present assessing the possibilities of refurbishing them for further use. If this approach succeeds, the present interim stage may well suffice for many years to come. However, if the cumulative payload that has to be established in orbit each year rises to a sufficiently high level, then it may be more reasonable to embark on the development of either recoverable winged rocket boosters which, after refurbishing, can be used again, or even the more ambitious so-called aerospace plane. This aerospace plane would be designed for the first stage to be flown back to base by a manned crew after it had achieved its appropriate stage velocity and separated from the remaining upper stages–which would carry on as at present to inject the payload into orbit. Although it would have to be designed from scratch, the aerospace plane would be tantamount to putting wings on the first stage of the Saturn V which is, after all, the largest and most expensive of the three rocket stages.

The development of a completely recoverable vehicle–spacecraft and launcher–to land on an airstrip is undoubtedly an expensive proposition to contemplate and figures of tens of billions of dollars are being recognised as necessary for such an undertaking. However, since space travel will be with us for ever and is no temporary excursion, a time will be reached when it will be more appropriate to have completely recoverable vehicles. The question is more one of when, rather than if, such devices will be developed.

Coming back to the manned space station, the crew will be a mixture of scientists and technologists, as well as the basic astronauts whose task will be to 'fly' the spacecraft. Already in both the Russian and American space programmes we have seen civilian scientists as members of spacecraft crews and, while at present there are obvious basic physical and health requirements for a man or woman going into orbit, it should be remembered that the American astronauts even now are typically around the age of forty. With the refinements that can be built into the systems of the future, there seems to be no basic reason why any healthy individual of a much greater age should not venture into space. Remembering also that a lot of work has already been carried out on hypersonic transports (which in many ways are really suborbital very fast supersonic airliners), the problem of a satisfactory environment for the more fragile humans will have to be met and overcome. While in a few decades time a journey into Earth orbit will not exactly be considered commonplace, it will certainly not be viewed with the drama and glamour that the pioneering flights of today justifiably command.

We can, therefore, see a pattern emerging for future space activities. There will be modular-built spacecraft of various sizes–Space Stations, Orbiting Laboratories, Workshops and so on–orbiting the Earth at altitudes from 300 to 22,300 miles. At the latter height a spacecraft is said to be in a geo-stationary orbit, because the period is 24 hours and equal to that of the Earth's rotation, so that the satellite remains fixed relative to the Earth's surface.

At the same time there will be various manned excursions to the Moon as a 'regular run' and within the next 10–15 years a Lunar scientific colony will probably be established. We shall also see the beginnings of the more adventurous operations in which manned interplanetary probes will voyage to Mars, Venus and beyond. The start for this development may well be in the next decade ready for operation in the early 1980s.

The allocations of funds between the direct application of space (which will certainly become predominant in the Earth-orbiting stations) and the basic scientific research (which for many years must remain the predominant criterion for the deeper space missions) have yet to be finally sorted out. However, the trends are fairly well established and it can be seen that the Apollo Project has played a fundamental part in blazing the trail for these further endeavours. Even manufacturing in Earth orbit is being examined, for it is possible to do quite remarkable things in the weightless conditions of space. For instance, perfectly spherical spheres for ball-bearings could be manufactured by allowing drops of molten metal to cool in the weightless environment, the sphere being the natural shape produced by surface tension. Foam steel could also be made by mixing air and molten steel into a homogeneous froth which, on cooling, would result in a honeycomb structure as light as balsa wood but with the strength of steel. Approval has, indeed, recently been given to NASA go carry out five such experiments in Earth-orbiting missions, one of which is the growth of crystals in weightless conditions. Soon we shall see whether these and other industrial techniques will add even further to the justification of such activity in outer space. This word 'justification' leads us finally to examine the question: 'Yes, but what is the real reason for and value of sending the Apollo crews to the Moon?'

The first thing to say is that it would be quite wrong to seek a short-term justification in terms of financial rewards covering the cost of the initial enterprise. The Apollo Project represents the visible peak of the most advanced technological challenge that the world has experienced so far and, of course, part of the value to the United States from this enterprise can be measured in prestige. It is also a fact that a major challenge like this stimulates technology on a broad front and produces usable equipment in other fields which is also of great importance. However, just below this visible peak there are two major divisions of activity which are perhaps the main reasons why it is right to engage in this project, and both of these are not always so visible to the population at large–rather like the hidden part of an iceberg.

The first of these activities is concerned with that which is learned by way of engineering techniques and technology in general, as a result of the very diverse work on components, sub-systems and systems, the standards for which are stretched to the limit in many instances to meet the stringent specifications such a space programme calls for. There are 22,000 companies working on this project alone in the United States and it must be obvious that this major challenge has produced a broad 'spin-off' or 'fall-out' which has been applied already to many varied techniques in other fields, even percolating into the consumer goods industries.

The other broad division of activity is in the direct *use* in other ways of the actual equipment that has been developed for the Apollo programme. For example, the direct use of the Apollo Command Module and its launch vehicles will be realised when they are applied to creating and maintaining the manned space stations which have already been mentioned. The components and sub-systems developed for Apollo are already being used directly on unmanned spacecraft and will be so used to an increasing extent in support of the direct application of satellites for communications, continuous meteorological surveying, earth resource operations (observing from space the natural resources of the Earth) and the like, all of which will help us better to know, to control and conserve our resources and environment.

Everything done by the Apollo Project, must, therefore, be considered on the credit side in the whole field of space. Much could have been done without the challenge of Apollo, as witness the efforts of some other countries, but it would not have happened with the same effectiveness or to the same extent without the spur of the tangible goal of creating a space capability by landing men on the Moon.

At all events we are still only beginning to scratch, as it were, the 'inner surface of outer space' in terms of exploration and exploitation, and no praise can be too high for the courage, purpose and ability of the Apollo crews, and of the men who designed and built their spacecraft, launching rockets and ground equipment, or planned and managed all that goes with the vast complexities of space travel. They are all helping to shape Man's destiny and changing the lives of us all in one way or another.

Index

INDEX

INDEX